Welding
Practice

Welding Practice

Brian D. Smith

Tech Eng, Tech WeldI, MITD, CGIA
Registered Welding Education Technician

ARNOLD

A member of the Hodder Headline Group
LONDON • SYDNEY • AUCKLAND

Acknowledgements

The author expresses his thanks to the following organisations for their support and their supply of information and illustrations:

Air Products PLC
Arc Speed Services, Derbyshire
Arco Ltd: safety signs, inside back cover
The British Oxygen Company (BOC), Guildford and Derby: gas cylinder identification chart, inside back cover and Figure 1.1
The British Standards Institute, Milton Keynes
CENTRA, Manchester
Chubb Fire Ltd: fire extinguishers, inside back cover
East Midlands Further education Council, Nottingham
ESAB Group (UK) Ltd, Waltham Cross
Gas Control Equipment, Skelmerdale
The Health and Safety Executive
Migatronic Welding Equipment Ltd, Loughborough
C. S. Milne Ltd, Leicester
Trueweld, Derby
The Welding Institute, Abington, Cambridge

Special thanks to Mr Len Gourd, BSc FWeldI for his continuing support.

Every possible effort has been made to trace copyright holders. Any rights not acknowledged here will be acknowledged in subsequent printings if notice is given to the publishers.

First published in Great Britain 1996 by
Arnold, a member of the Hodder Headline Group,
338 Euston Road, London NW1 3BH

British Library Cataloguing in Publication Data
A catalogue record for this book is available from the British Library

ISBN 0 340 61406 4

Produced by Gray Publishing, Tunbridge Wells
Printed and bound in Great Britain by J. W. Arrowsmith Ltd, Bristol

Contents

1
Underpinning information

Arc welding safety

Under the Health and Safety at Work Act 1974 welders have a responsibility to take reasonable care for their safety and that of others by cooperating with safety requirements. Welding and the welding environment can create many hazards, but can be carried out quite safely if you play your part and observe some basic safety rules. There are a number of factors to be considered; a few of these are listed below.

Personal protection

Personal protection for the welder is most important, apart from the most obvious dangers of burns, stray sparks and falling objects, the arc gives off ultraviolet and infrared rays. These will affect the skin and eyes much in the same way as long periods of exposure to the sun. Suitable protection must be worn at all times to guard against these and other dangerous occurrences.

Protective clothing comes in the form of gauntlets, leather coats, aprons, sleeves, spats and capes.

The eyes should always be protected by a shade of filter glass suitable for the welding operation and amperage being used, or a clear glass for the chipping of slag. Table 1.1 details the recommended filter glasses.

Protection for the feet in the form of safety boots with a toecap is an essential part of the welder's protective clothing. Boots protect the

Table 1.1   Recommended filter glasses for welding

Filter glasses for manual metal arc welding		
8–9	EWF	up to 100 amps
10–11	EWF	100–300 amps
12–13–14	EWF	over 300 amps
Filter glasses for oxyfuel gas welding:		
3	GWF	Aluminium and alloys
4	GWF	Brazing and bronzing welding
5	GWF	Copper and alloys
6	GWF	Thick plate and pipe

Filter glasses are usually protected on the outside by a separate, clear plastic cover lens. This will give protection from spatter particles and prolong the life of the filter lens.

toes from damage by falling objects, while also giving the best protection against sparks entering the footwear. Some boots have a flap up the front, others may be one piece with high legs giving good protection to the foot and leg. Low shoes or trainers are not the ideal footwear for the welder.

Many types of eye protection are available in the form of spectacles, goggles, face visors, clear glasses for use with welding helmets/ shields, etc. The most important aspect when choosing any form of eye protection is that it must conform to British Standard (BS)2092. Eye protection must be worn for all cutting, grinding, chipping and welding operations.

Working at heights

In your profession as a welder it is possible that you may have to work off scaffolding, trestles, elevated platforms and so on. In most instances these elevated platforms will be erected or operated by qualified people eliminating the possible risk of collapse.

As a welder working from such platforms, extra care must be taken to avoid even the slightest risk of electric shock. Although the effects of the shock may be dismissed, it could lead to loss of balance and falling with serious or fatal results.

Remember **always** take extra care when working at heights.

Responsibilities of the welder

Under the Health and Safety at Work Act 1974, individuals are held responsible for their own safety as well as that of colleagues and those working in close proximity.

Always ensure you have taken all reasonable, practical precautions to avoid the risk of accidents and fires, for example:

a) good housekeeping; keep a tidy work area
b) wear protective clothing on feet, body and eyes
c) screening of workstation to protect passers-by
d) use and position extraction equipment correctly
e) carry out safety checks on equipment
f) do not tamper with safety posters or signs.

These are just a few points to consider before work commences. How many more can you think of?

Welding fumes

Welding fumes are virtually impossible to eliminate from the welding process. However, fumes can be rendered harmless by observing the following simple rules:

a) the welder positioning him/herself out of the fume path
b) using adequate ventilation either natural or mechanical – fans and extraction equipment, for instance
c) correct positioning of source of ventilation – some gases are heavier than air and will sink to the bottom of a confined area; in these cases extraction should be at a low level

d) placing the source of extraction as close to the point of welding as possible without disturbing the gas shield
e) siting the source of extraction to pull the fumes away from the welder
f) avoid welding on contaminated surfaces, such as oil, grease, paint, galvanised metal, etc.

Remember, fumes which include poisonous gases which asphyxiate are not always visible. Where it is not possible to use the above methods, then breathing apparatus may be required, supplied either from an airline or a personal body pack.

If in doubt ask!

Storage and handling of gas cylinders

The most common method of supplying gases used for welding is from cylinders. All gas cylinders should be treated with respect, handled carefully and stored in well-ventilated conditions. Never allow cylinders to come into contact with heat, contaminants containing oil and grease, and observe the following handling and storage conditions:

a) always keep cylinders, fittings and connections free from oil and grease
b) store cylinders upright.
c) transport cylinders upright.
d) store flammable and non-flammable gases separately
e) store full and empty cylinders separately
f) avoid cylinders coming into contact with heat
g) secure cylinders by 'chaining up' during storage, transport and use
h) open cylinder valves slowly
i) only use recommended leak-detection sprays
j) report damaged cylinders to the supplier
k) always check for leaks
l) switch off gas supply after use.

Gas cylinders find their way into many different environments from schools and colleges to heavy fabrication. Wherever gases are used the operator, supervisor and management should be aware of the dangers and familiar with the operating conditions, safety requirements, handling and storage conditions. Gas suppliers can provide information on safety, operating instructions, fumes and gas cylinder identification charts to assist users in operating safely (see Fig. 1.1 for a gas cylinder identification chart.).

Correct earthing

In modern welding workshops, all structural metalwork and metalwork forming the fabric of the building is normally connected to the bonded earth system of that building.

When welding on any steelwork, it is essential that the power return cable of the welding equipment is securely clamped to the workpiece, close to the point of welding. It must also be ensured that there is no break, or poor connections, which would interrupt the flow of current in the circuit, resulting in poor welding conditions or dangerous situations.

The power return cable is connected to the work by a clamp or bolting direct, ensuring that it makes a good electrical connection. It is important that at the point of contact the area is clean and free from rust, paint scale or other surface coatings which may result in a poor electrical contact.

Figure 1.2 shows a good circuit with clean, positive connections, the power return cable is

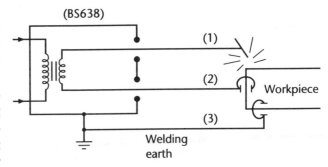

The proper arrangement
(1) Welding circuit isolated from earth at the. welding set, i.e. to BS638.
(2) Welding return lead.
(3) Separate workpiece earthing lead.

Figure 1.2   Correct earthing arrangement.

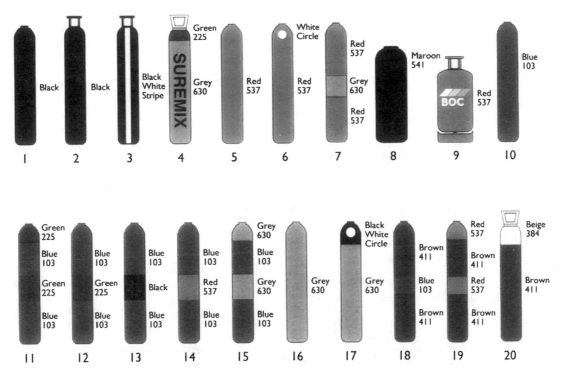

Figure 1.1   Gas cylinder identification chart (industrial). Key: (1) Oxygen. (2) Carbon dioxide/Suregas – vapour withdrawal. (3) Carbon dioxide – liquid withdrawal. (4) Carbon dioxide/nitrogen mixture – Suremix. (5) Hydrogen. (6) Hydrogen high purity. (7) Hydrogen and nitrogen. (8) Dissolved acetylene. (9) Propane. (10) Argon and high purity argon and Pureshield argon. (11) Argoshield. (12) Argon + CO$_2$ (Pureshield P3). (13) Argon + oxygen (Argonox). (14) Argon + hydrogen (Pureshield P1, P2, P4). (15) Argon + nitrogen (Pureshield P5). (16) Air. (17) Nitrogen (Oxygen free). (18) Helishield (H1, H2, H3, H5, H6, H7, H101). (19) Helishield (H4). (20) Balloon gas.

clamped close to the point of welding causing minimum resistance to the flow of current in the circuit.

If a break should occur in the return power cable above, the arc will extinguish, the system is then safe and the power can be switched off while the fault is located and rectified.

Fuses

Welding circuits are usually fitted with fuses which protect the circuit in the event of a sudden overload of current.

If a situation exists as in Fig. 1.3 where the power return cable is clamped to adjacent metalwork, or the workpiece comes into contact with the metal structure, the return path of the current will take the route of least resistance. This may be through cables of power tools or other electrical apparatus attached to, or in contact with, the work. If this occurs, it can cause cables attached to electric motors or power tools to burn out, resulting in sparks and fires. The result of this happening in a hazardous atmosphere or confined space can be catastrophic.

Remember

■ always ensure a good, clean, positive electrical connection

■ locate the power return cable clamp close to the point of welding
■ check cables for worn or damaged insulation and breakage
■ ensure cables are large enough for the amperage rating of equipment.

Fire prevention

When cutting and/or welding, sparks and flames are two of the many hazards of the trade. Working in many different environments you will encounter materials which will readily catch fire. It is your responsibility to ensure that all reasonable precautions are taken to prevent the outbreak of fire

■ always have a suitable fire extinguisher to hand
■ remove any flammable materials from the work area
■ check both above and below where you are working
■ let other people know where you are working
■ do not work over the top of your equipment
■ remove any canisters or containers
■ damp down wooden flooring or scaffold boards
■ wear protective clothing.

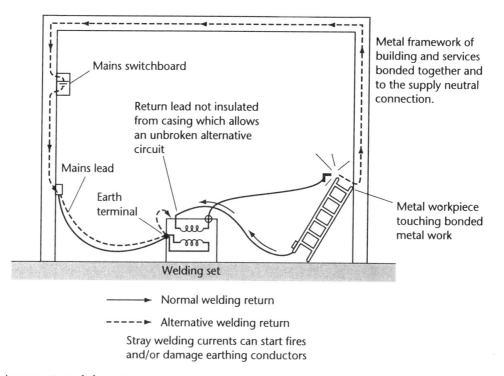

Figure 1.3   An incorrect and dangerous arrangement.

These are just a few points for you to think about. Always consider the working environment and ask yourself: **what could cause a fire? – remove it and prevent it from happening.**

Electrical hazards

The risk of electric shock is something which the welder should always be aware of, both to themselves and colleagues. Injury resulting from electric shock can be burns, loss of consciousness or death.

If you come across someone who you suspect has received an electric shock, **switch off** the supply immediately and **only** remove the injured person from contact by means of suitably insulated protection. For normal arc welding voltages, sufficient protection will be given by their clothing – provided it is dry. When freeing the injured person from the electrical source, it must be done with the use of a non-conducting material.

Where staff trained in first aid are available, inform them immediately so they are able to administer the correct treatment. If no-one is available, and the situation is serious, attempt mouth-to-mouth resuscitation while you send for help.

Remember that where there is a suspected risk of electric shock, due to working conditions or environment, adequate precautions must be taken.

However minor the incident may appear, always ensure that the injured person is seen by a doctor.

It is also advisable that those involved with electric arc welding be trained in the basic practice of mouth-to-mouth resuscitation.

Welding terminology

Welding, gas and arc, is a widely used process to join metals by fusion welding, brazing or bronze welding.

The terminology tends to vary somewhat from county to county, and only contact, conversation and experience can create an awareness of the wide range of names used.

This book has put together a list of terms based on the most widely used to describe the various features of the process. Table 1.2 gives terms relating to the gas and arc welding processes. Table 1.3 specifically relates to terms used in the thermal cutting of metals, and Table 1.4 deals with testing and examination.

Table 1.2 Terms relating to gas and arc welding

Term	Definition
Actual throat thickness	The perpendicular distance between two parallel lines joining the weld toes
Arc blow	A disturbance of a DC welding arc caused by magnetic fields set up in the work
Arc voltage	The voltage between electrodes or between an electrode and the work during welding
Backfire	Retrogression of the flame into the blowpipe neck or body with rapid self-extinction
Back-step sequence	A welding sequence in which short lengths of weld are deposited in a direction opposite to the direction of progress along the joint to produce a continuous or intermittent weld
Backing strip (backing ring pipe)	A piece of metal placed at a root and penetrated by weld metal. It may remain as part of the joint or removed by machining or other means
Backing bar (backing ring pipe)	A piece of metal or other material placed at a root and used to control the root penetration bead
Burn back	Fusing of the electrode wire to the current contact tube in any form of automatic or semi-automatic metal-arc welding process
Burn off rate	The linear rate of consumption of a consumable electrode in any consumable electrode process
Burn through (melt through)	A localised collapse of the molten pool due to excess localised heating, resulting in a hole in the underlying weld run or parent metal
Cutting/welding torch	A device for mixing and burning gases to produce a flame for welding, brazing, bronze welding, cutting, heating and similar operations

Table 1.2   Continued

CO₂ flux cored welding (cord wire welding)	Metal arc welding in which a flux cored electrode is deposited under a shield of carbon dioxide
CO_2 welding	Metal arc welding using a continuous bare wire electrode. The arc and molten pool being shielded by carbon dioxide shielding gas
Concave fillet weld	A fillet weld in which the weld face is concave (curved inwards)
Cone	The inner part of a flame adjacent to the nozzle orifice – known as the inner cone (oxyfuel gas welding)
Continuous weld	A weld along the entire length of a joint
Convex fillet weld	A fillet weld in which the weld face is convex (curved outwards)
Covered filler metal	A filler metal having an outer covering of flux can be in the form of a continuous covering or contained in indentations along its length
Crack	Discontinuity in the welded joints. Cracks may be longitudinal, transverse, edge, crater, centre line, fusion zone underbead, weld metal or parent metal
Crater pipe	A depression due to shrinkage at the end of a run (crater) where the source of heat was removed
Deposited metal	Metal after it becomes part of a weld or joint
Dip transfer	A method of metal-arc welding in which fused particles of the electrode wire in contact with the molten pool are detached from the electrode in rapid succession by the short circuit current, which develops every time the wire touches the molten pool
Dual shield welding	Semi-automatic welding using a flux covered wire and a shielding gas
Excess penetration bead	Metal protruding through the root of a weld made from one side only in excess of the stated limits
Feather	The carbon-rich zone, visible in a flame, extending around and beyond the cone when there is an excess of fuel gas
Fillet weld	A fusion weld, other than a butt or edge weld, which is approximately triangular in cross-section
Filler metal (filler wire or filler rod)	Metal added during welding
Flame snap-out	Retrogression of the flame beyond the blowpipe body into the hose, with possible subsequent explosion
Flashback arrester	A safety device fitted in the oxygen and fuel gas system to prevent any flashback reaching the gas supply
Flux	Material used during welding to prevent atmospheric oxidation and to reduce impurities or float them to the surface. Can also have a cleansing action on the surfaces to be joined
Fusion penetration	The depth to which the parent metal has been melted into the fusion faces
Fusion welding	Joining together to form a union between metals in a molten state without the application of pressure
Fusion zone	An area of the parent metal at the fusion face which is melted to form part of the weld
Gas economiser	A device designed for temporarily cutting off the supply of gas to the welding equipment. A pilot jet may be fitted for relighting
Gas envelope	The gas surrounding the inner cone of an oxyfuel gas flame
Gas pore (gas cavity)	A cavity formed by entrapped gas during the solidification of molten metal
Gas regulator	A device for attachment to a gas cylinder or pipeline for reducing and regulating the cylinder or line pressure to the working pressure required
Globular transfer	Metal transfer which takes place as large globules transferred from the electrode to the weld area

Table 1.2   Continued

Heat-affected zone (HAZ)	The part of the parent metal which is metallurgically affected by the heat of welding or thermal cutting
Hose protector	A small non-return valve fitted to the torch end of a hose to resist the force of a flashback
Included angle (angle of preparation)	The angle between the fusion faces of parts to be welded
Inclusion	Slag or foreign matter entrapped during welding. The defect is usually more irregular in shape than a gas pore
Incomplete root penetration	Failure of weld metal to extend into the root of a joint
Incompletely filled groove	A continuous or intermittent channel in the surface of a weld due to insufficient weld metal. The channel may be in the centre or along one or both edges of the weld
Lack of fusion	Lack of union melting in a weld either: **a)** between weld metal and parent metal, **b)** between parent metal and parent metal or **c)** between weld metal and weld metal
Leg	The distance from the weld toe to the root of the joint (size of fillet weld)
Leftward welding (forward welding)	A gas welding technique in which the flame is directed towards the unwelded part and the filler rod, when used, is directed towards the welded part of the joint
Metal arc welding	Arc welding using a consumable electrode
Metal transfer	The transfer of metal across the arc from a consumable electrode to the molten pool
MIG/MAG welding	Metal arc welding using continuous bare wire electrodes, the arc and the molten pool being shielded by an inert or mixed shielding gas
Parent metal	Metal to be joined
Penetration bead	Weld metal protruding through the root of a weld made from one side only
Plug weld	A weld made by filling a hole in one component of a workpiece to join it to the surface of an overlapping component
Porosity (surface or internal)	A group of gas pores
Open circuit voltage	The voltage between two output terminals which are carrying no current
Open arc welding	Arc welding in which the arc is visible
Overlap	An imperfection at a toe or a root of a weld caused by metal flowing on to the surface of the parent metal without fusion
Residual welding stress	Stress remaining in a metal part or structure after welding has taken place
Rightward welding (backward welding)	A gas welding technique in which the flame is directed towards the welded part and the filler rod is directed towards the unwelded part of the joint
Root (of weld)	The zone on the side of the first run farthest from the welder
Root face	The portion of a fusion face on the joint preparation which is not bevelled or grooved
Run	The metal melted or deposited during one pass of an electrode, torch or blowpipe (also known as a pass or bead)
Run-off-plate(s)	A piece, or pieces, of metal so placed as to enable the full section of weld metal to be obtained at the end of a joint avoiding craters
Run-on-plate(s)	A piece or pieces of metal so placed as to enable the full section of weld metal to be obtained at the beginning of a joint avoiding start problems
Sealing run (backing run)	The final run deposited on the root side of a fusion weld
Shrinkage groove	A shallow groove caused by contraction of the metal along each side of a penetration bead
Skip sequence	A welding sequence in which short lengths of run are spaced in planned positions in order to produce a continuous or intermittent weld

Table 1.2 Continued

Slag-trap	A feature in a joint or joint preparation which may lead to the entrapment of slag
Slot weld	A weld made between two overlapping components by depositing a fillet weld round the periphery of a hole in one component so as to join it to the other component
Spray transfer	Metal transfer which takes place as a steam of small droplets transfers from the electrode to the weld area
Stack cutting	The thermal cutting of a stack of plates usually clamped together
Staggered intermittent weld	An intermittent weld on each side of a joint arranged so that the welds lie opposite to one another
Striking voltage	The minimum voltage required to strike an arc
Submerged arc welding	Metal arc welding in which a bare wire or electrodes are used, the arc or arcs are covered by a flux, some of which fuses to form a removable slag on the weld. Some flux is recovered
Sustained backfire	Retrogression of the flame into the blowpipe, the flame remaining slight. 'Popping' or 'squealing' with a small pointed flame coming from the nozzle or as a rapid series of minor explosions
TIG welding (inert gas tungsten arc welding)	Inert gas welding using a non-consumable electrode of pure or activated tungsten
Touch welding	Metal arc welding using a covered electrode. The covering is kept in contact with the parent metal during welding
Toe	The boundary between a weld face or root and the parent metal, or between a weld face and any underlying welds
Tungsten inclusion	An inclusion of tungsten in the weld from the electrode in TIG welding process
Two-stage regulator	A gas regulator in which the cylinder or line pressure is reduced to the working pressure in two stages
Undercut	An irregular groove at a toe of a run in the parent metal at root, face or in previously deposited weld metal
Weld	A joint between pieces of metal made liquid by heat, or by pressure, or by both. A filler metal may or may not be added
Weld junction	The boundary at the extent of melting and the heat-affected zone
Weld metal	The metal melted during the making of a weld and retained in the weld joint
Weld zone	The area of parent metal affected by the weld deposit
Welding procedure	A detailed list of actions to be followed during the production of a weld. The list will include materials, consumable, amps, volts, gases, joint details, etc.
Welding sequence	The order and direction in which welds are deposited in the joint
Welding technique	The manner in which the operator controls the electrode or blowpipe during welding
Worm-hole	An elongated or tubular cavity formed by entrapped gas during solidification of molten metal

Table 1.3 Terms relating to thermal cutting

Term	Definition
Air-arc cutting	Thermal cutting using an arc for melting the metal and a stream of air to remove the molten metal
Cutting electrode (gouging electrode)	An electrode with a covering that aids the production of such an arc that molten metal is blown away to produce a groove or cut
Cutting oxygen	Oxygen used at a pressure suitable for the cutting of steels
Drag	The projected distance between the two ends of a drag line
Drag lines	Serrations left on the face of a cut edge made thermal by cutting
Flame cutting	Oxygen cutting in which the appropriate part of the material to be cut is raised to ignition temperature by an oxyfuel gas flame
Flame gouging	A method of surface shaping and dressing of metal by flame cutting
Flashback arrester	A safety device fitted in the oxygen and fuel gas system to prevent any flashback reaching the gas supply
Gas regulator	A device for attachment to a gas cylinder or pipeline for reducing and regulating the cylinder or line pressure to the working pressure required
Hose protector	A small non-return valve fitted to the torch end of a hose to resist the force of a flashback
Kerf (kerf width)	The width of cut left after metal has been removed by thermal cutting
Metal arc cutting (gouging)	Thermal cutting by melting using the heat of an arc between a metal electrode and the metal to be cut
Oxygen arc cutting (oxy-arc cutting)	Thermal cutting in which the ignition temperature is produced by an electric arc, and cutting oxygen is conveyed through the centre of an electrode, which is consumed in the process
Oxygen lance	A steel tube, consumed during cutting, through which oxygen passes, for the cutting and boring of holes
Oxygen lancing (thermic lancing)	Thermal cutting process in which an oxygen lance is used
Packed lance	An oxygen lance packed with steel rods or wires
Preheating oxygen	Oxygen used at a suitable pressure in conjunction with fuel gas for raising the metal to be cut to ignition temperature
Thermal cutting	The parting or shaping of materials by the application of heat with or without a stream of cutting oxygen
Two-stage regulator	A gas regulator in which the cylinder or line pressure is reduced to the working pressure in two stages

Table 1.4 Terms relating to testing and examination

Term	Definition
All-weld metal test specimen	A test specimen that is composed wholly of weld metal
Coupon plate (test coupon)	A test piece made by adding plates to the end of a joint to give an extension of the weld for testing purposes
Cruciform testpiece	A flat plate to which two other flat plates or two bars are welded at right angles and on the same axis
Destructive testing	Tests carried out; where the weld is subject to some form of mechanical testing

Table 1.4 Continued

Face bend test	A bend test specimen in which the face of the weld is in tension
Free bend test	A bend test made without using a former
Guided bend test	A bend test made by bending the specimen round a specified former
Nick break test	A fracture test in which a specimen is broken from a notch The notch is located to determine the point of fracture. Used for internal examination of weld
Non-destructive testing	Tests carried out on a weld without causing mechanical damage
Reverse bend test (root bend test)	A bend test specimen in which the root of the weld is in tension
Side bend test	A bend test in which the through thickness of a weld section is in tension
Test piece	Components welded together in accordance with a specified welding procedure, or a portion of a welded joint detached from a structure for test
Test specimen (test coupon)	A portion detached from the weld containing the whole weld area, and prepared as required for testing
Tongue bend test specimen	A portion cut in two straight lengths of pipe joined by a butt weld to produce a tongue containing a portion of the weld. The cuts are made parallel to the axis of the pipes and the weld is tested by bending the tongue round a former

Weld symbols (BS499 part 2 1980)

To avoid confusion and limit the amount of written instruction on fabrication drawings, a code of symbols has been devised and compiled into a booklet to form BS499 part 2, 1980.

BS499 is a comprehensive document which lists all the weld symbols and their graphical shape, the symbols being common to all welding processes. This section will look at the symbols applied to butt and fillet welds, although the basic rules apply in all situations and it is these basic rules we need to consider before moving on.

The whole of the weld symbol is made up of:

a) the **arrow** line
b) the **reference** line
c) the **weld** symbol.

The arrow line is to indicate the position of the weld at the joint. The side nearest to the arrow is known as the **arrow side**. The side farthest away from the arrow is known as the **other side**. It is important to remember that where only one side of the joint is a prepared edge, the arrow will always point toward the prepared edge.

The reference line should always be hori-

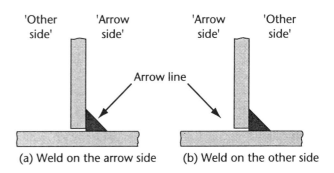

(a) Weld on the arrow side   (b) Weld on the other side

Figure 1.4   (a) Weld on the 'arrow side'. (b) Weld on the 'other side'.

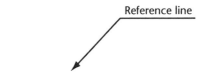

Figure 1.5   A reference line.

zontal, e.g. parallel to the bottom edge of the drawing, with one end touching the arrow line to form an angle between them (Fig. 1.5). The reference line holds all the information about the joint and the weld.

The symbol or symbols appear above or below the reference line. If the symbol is **above** the line the weld is on the **other** side. If the symbol is **below** the line the weld is on the **arrow** side (see Fig. 1.6).

In some instances it will be necessary to use a combination of symbols both above and

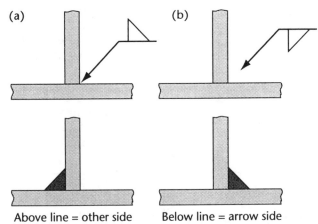

Above line = other side   Below line = arrow side

Figure 1.6 (a) Symbol above line means weld 'other side'. (b) Symbol below line means weld 'arrow side'.

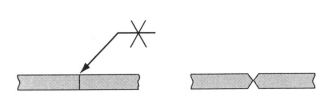

Figure 1.7 Symbols above *and* below line mean work on both 'arrow side' and 'other side'.

below the reference line. An example of this is shown in Fig. 1.7.

Identify the type of weld

Figure 1.8 shows two types of weld. The following symbols also signify a butt weld:

μ   = single 'J' butt weld
ᴗ   = single 'U' butt weld
ν   = single bevel butt weld.

Identify the joint preparations show in Fig. 1.9.

Where weld sizes are indicated, for example, fillet weld leg length and throat thickness are shown on the left-hand side of the symbol.

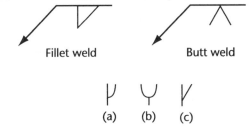

Fillet weld   Butt weld

μ   ᴗ   ν
(a) (b) (c)

Figure 1.8 Symbols showing type of weld.

Symbol	Type of joint	Preparation
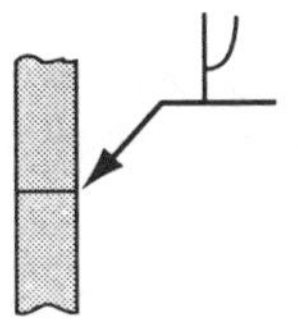	Single 'J' butt other side bottom plate prepared	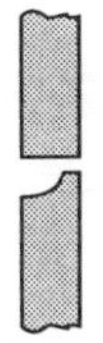
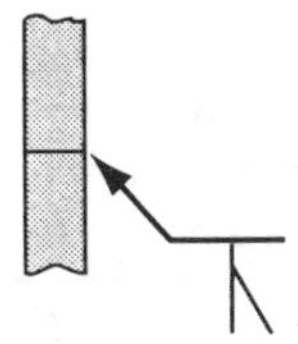	Simple bevel butt arrow side top plate prepared	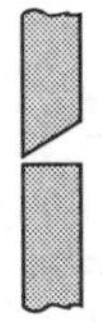
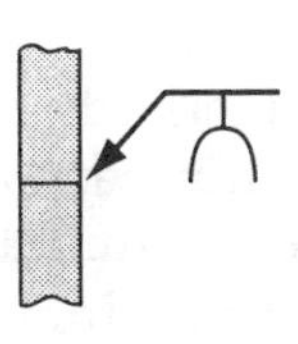	Single 'U' butt arrow side	
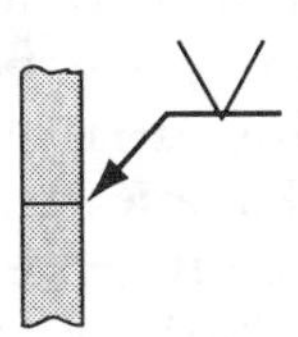	Single 'V' butt other side both plates prepared	

Symbol	Type of joint	Preparation
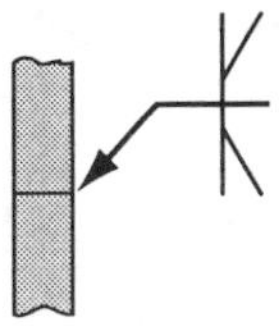	Double bevell butt both sides bottom plate prepared	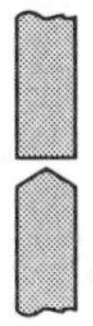
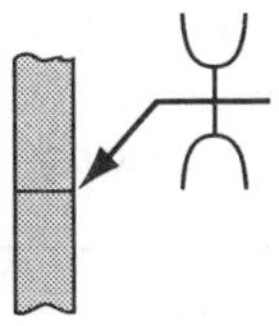	Double 'U' butt both sides both plates prepared	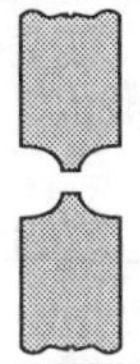
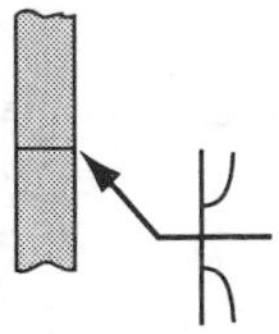	Double 'J' butt both sides top plate prepared	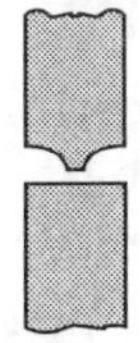
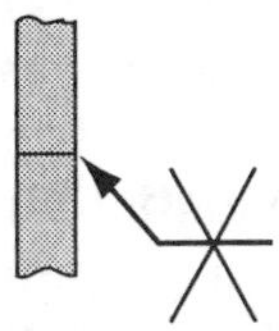	Double 'V' butt both sides both plates prepared	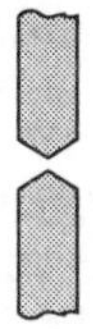

Figure 1.9 Where the symbol is made up of a vertical line this will always be on the left-hand side of the symbol. These are the basic rules applied to understanding weld symbols.

When it is necessary to show the throat thickness (*a*) and the leg length (*b*) on the same symbol, the throat thickness will be preceded by the letter *a* and the leg length by the letter *b* (see Fig. 1.10 – this indicates a throat thickness of 9 mm and a leg length of 12 mm).

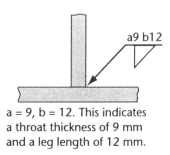

a = 9, b = 12. This indicates
a throat thickness of 9 mm
and a leg length of 12 mm.

Figure 1.10   Weld dimensions.

Weld run length dimensions are shown on the right-hand side of the symbol. In the case of an intermittent fillet weld the dimensions appearing to the right-hand side of the symbol would be:

■ the number of the welds, represented by the letter *n*
■ the length of the welds, represented by the letter *l*
■ the space between the welds, represented by the letter *e* (see Fig. 1.11).

Joint	Symbol	Description
	8 ⊿ 16 x 50(50) n x e(e)	8 mm leg length, fillet weld n = 16 welds l = 50 mm long e = 50 mm spacing

Figure 1.11   Number, length and spacing of welds.

Additional symbols can be used to give advice on the finished weld surface shape, for example:

— = flush finish
⌢ = convex finish
⌣ = concave finish (see Fig. 1.12).

The symbol does not indicate how this shape will be achieved.

In the case of the butt welds, unless otherwise stated, it is assumed that the joint will be a full penetration butt joint. In the case of par-

Symbol	Shape	Description
6		6 mm leg length fillet weld concave finish
		Full penetration butt weld flush finish
10		10 mm leg length fillet weld convex finish

Figure 1.12   Shape of finished weld.

tial penetration butt welds the depth of penetration (e.g. the distance from the plate surface to the bottom of the penetration), would be indicated on the left-hand side of the symbol. In the case of double-sided welds the depth of penetration should be given for both welds. In BS499 this dimension is indicated by the letter *s* (see Fig. 1.13).

Symbol	Shape	Description
4	s = 4mm	Single 'V' butt weld arrow side 4 mm penetration
6 4	s = 4mm s = 6mm	Double 'V' butt weld 4 mm penetration, arrow side 6 mm penetration, other side

Figure 1.13   Weld penetration.

Welds to be carried out on site, methods of examination and process used is all information which you could find on the weld symbol, descriptions of which are given in Fig. 1.14.

This has been a brief introduction to weld symbols but one which will form the basis for the interpretation of all the ones you are likely to encounter. Most weld symbol shapes are designed to look like the weld shape it represents, so by remembering these few basic rules it will be possible to understand most weld symbols used.

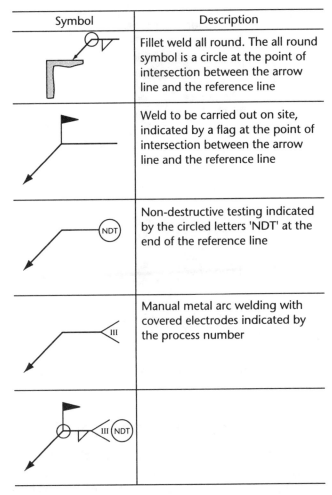

Symbol	Description
	Fillet weld all round. The all round symbol is a circle at the point of intersection between the arrow line and the reference line
	Weld to be carried out on site, indicated by a flag at the point of intersection between the arrow line and the reference line
NDT	Non-destructive testing indicated by the circled letters 'NDT' at the end of the reference line
III	Manual metal arc welding with covered electrodes indicated by the process number
III NDT	

Figure 1.14 Other symbols used on weld diagrams.

The reason for British Standards is so that in the course of the work reference can be made to them. Further information on weld symbols can be found in BS499 Welding Terms and Symbols, part 2 Symbols for Welding.

Types of joints

Preparing the joint to accept the weld is sometimes a very complex exercise as in the case of thick plate butt welding by conventional techniques, e.g. MMA, TIG, MIG/MAG welding processes. Square edge butt preparation would be the ideal situation as this is probably the cheapest to produce, but the technique is limited to the thinner metals.

Machining or flame-cutting plate edges is an expensive exercise and, after all, when the metal has been removed it has to be filled up again with weld metal. The main reason for

joint preparation is to allow for access to the root of the joint in order to achieve penetration and produce a full strength joint.

All the joint profiles shown below are suitable for MMA, MIG/MAG and TIG processes. Figure 1.15 shows examples of common joint types, names, profiles and dimensions.

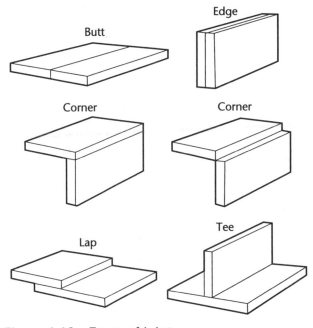

Figure 1.15 Types of joint.

Square edge preparations can be used for plate thicknesses of 1.6 mm up to and including 6 mm with a relevant gap setting, vertical up or vertical down technique (see Fig. 1.16).

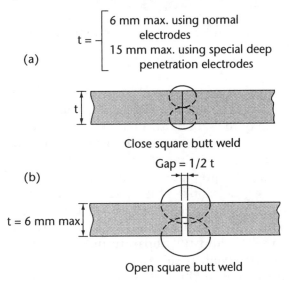

Figure 1.16 (a) Close square butt weld. (b) Open square butt weld.

Single 'V' preparations are used when access is needed to the root from one side only, through the full thickness of the plate. Suitable for thicknesses over 6 mm although this preparation becomes uneconomical above 15–20 mm.

Double 'V' preparations are used where access is available from both sides of the joint. When using double 'V' preparations it is possible to control distortion to some degree by balancing the weld deposits either side of the plate. Control of the penetration bead is not so critical as when the first run has been deposited in the bottom of the 'V' in the first side; the root can be ground back before depositing the first run on the other side.

Examples of single and double 'V' preparations are shown in Fig. 1.17.

(a)

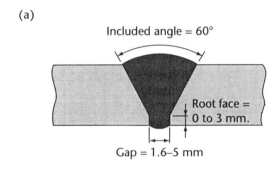

(b)

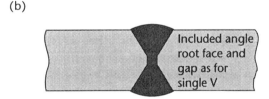

Figure 1.17 (a) Single 'V' butt weld. (b) Double 'V' butt weld.

Single and double bevel butt joints are used where a full penetration weld is required, but there may be a slight change in thickness between the two plates or the joint configuration may be a 'T'-shape (see Fig. 1.18).

'U' and 'J' preparations are generally used on thicker sections. This type of preparation tends to have steeper sides to the groove reducing the width at the top of the groove. Both the 'U' and 'J' preparations can be used where one or both sides of the joint are prepared (see Fig. 1.19).

Economical joint preparation can result in reducing production costs by saving on con-

(a)

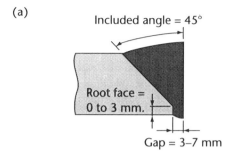

(b)

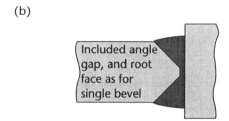

Figure 1.18 (a) Single bevel butt weld. (b) Double bevel butt weld.

sumables, preparation time, welding time, operator fatigue, reduction of fumes, etc.

Features of a welded joint

Knowledge of welding techniques, the process being used and the skill of the individual are all qualities needed by the welder, applying the skill and knowledge will achieve acceptance levels required by particular codes and standards. In order to apply the various acceptance levels to the actual welded joint it is important that the welder understands the different areas of the weld zone. This also plays an important role when describing specific details or requirements of the weld or joint to an inspector, colleague or workmate.

Figure 1.20 shows the terminology to be used when referring to specific sections of a butt and fillet weld.

Weld size

As mentioned earlier, in the case of a fillet weld the size is usually denoted by the leg length. If no specific dimension is stated it is a rule of thumb to make the weld size equal to the thinnest member in the joint. The size of a butt weld on the other hand is usually determined by the thickness of the metal being joined as shown in Fig. 1.21.

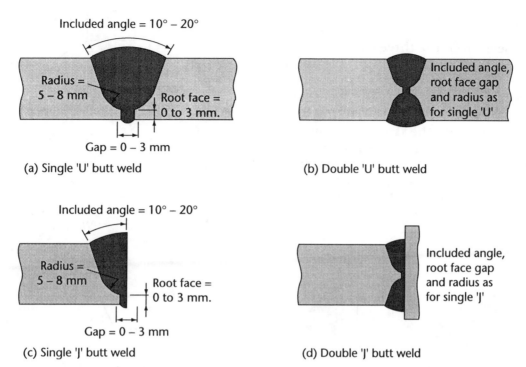

(a) Single 'U' butt weld

(b) Double 'U' butt weld

(c) Single 'J' butt weld

(d) Double 'J' butt weld

Figure 1.19 (a) Single 'U' butt weld. (b) Double 'U' butt weld. (c) Single 'J' butt weld. (d) Double 'J' butt weld.

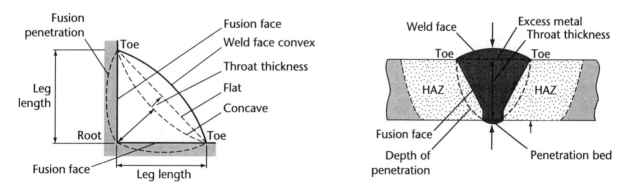

Figure 1.20 Features of a welded joint.

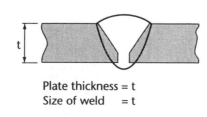

Plate thickness = t
Size of weld   = t

Figure 1.21 Thickness of a butt weld.

Fillet weld profiles

As can be seen from the three examples shown in Fig. 1.22, the finished weld profile and the angle between the joint faces can affect the throat thickness of the weld. The dimension *a* design throat thickness is used by the design engineer to calculate the strength of the welded joint and hence that of the total structure to ensure it will stand up to service conditions. This dimension is not easy to determine or to control by the welder during the welding operation.

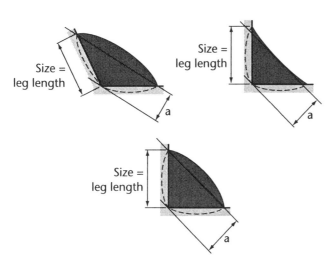

Figure 1.22   Throat thickness and leg length of a fillet weld.

To enable the design engineer to achieve the calculated throat thickness required details are conveyed to the welder by using the leg length dimension that will result in achieving the desired throat thickness.

By using this method the welder has some visual dimension to work to and thus achieves the previously calculated design throat thickness.

Features of the weld

See Fig. 1.23.

Features of the joint

See Fig. 1.24.

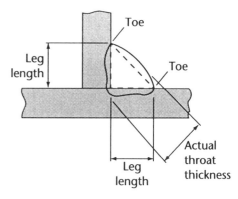

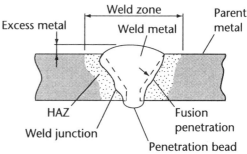

Figure 1.23   Features of the weld.

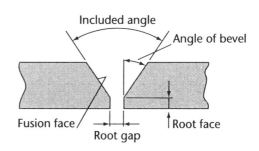

Figure 1.24   Features of the joint.

Distortion control

When the temperature of a metal bar is raised dimensional changes take place. This is referred to as **expansion** (Fig. 1.25). When allowed to cool without restraint (freely), the metal bar will return to its original size.

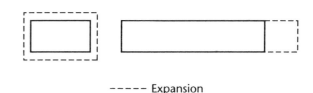

----- Expansion

Figure 1.25   Expansion of metals on heating.

This section will discuss some of the effects of heat on metal. The heat source in this case will be supplied by the welding process, e.g. MMA, MIG/MAG, TIG or O/A. The metal bar will be replaced by plates which will be joined by a weld to form the joint.

The temperature of the welding flame or arc is approximately 3200 or 4000°C, respectively. When these temperatures are applied to the plates to be joined, the metal will expand and, on cooling, will want to return to its original size. As we can see in Fig. 1.26 expansion and contraction take place both along the length of the bar (longitudinal) and also across the bar (transverse). This will apply to the heated plates which are joined together by the weld.

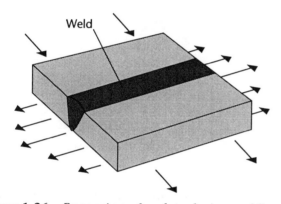

Figure 1.26   Expansion of a plate during welding.

As we can see in Fig. 1.26, as the weld is deposited and the temperature of the plates increases, expansion will take place. The hottest part of the plates will be along the weld joint itself, gradually reducing towards the plate edges. When the plates cool down, maximum cooling, and hence contraction, will take place in the weld area – (the area having reached the highest temperature).

This will, by the effect of length contraction, cause the plates to bend along their length – upwards towards the site of welding (Fig. 1.27).

On a cross-section of the weld, there is a large proportion of molten metal deposited above the centre line of the plate thickness (Fig. 1.28) and therefore more contraction takes place toward the top face, which causes the plates to bend upwards. This also has an effect on the transverse distortion.

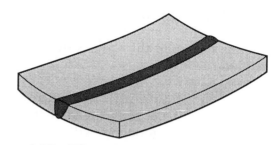

Figure 1.27   Effect of cooling once weld is complete.

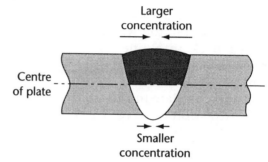

Figure 1.28   Uneven contraction between top and bottom surface of a plate.

The shaded area in Fig. 1.28 shows the amount of molten metal deposited above the centre line of the plate thickness – resulting in maximum contraction taking place along the top edge of the plate. This causes the plates to hinge upward from the root of the weld.

In the processes under discussion, it is not possible to **eliminate** the causes of distortion; it is, however, possible to **prevent** the distortion from taking place – and this is our next topic for consideration.

To overcome the problems of distortion, two methods are prevalent:

- one is **control** and
- the second is **correction**.

In all cases it is far better to **control** rather than **correct**. To quote an old adage 'prevention is better than cure'!

Methods of controlling distortion

Tack welding is one of the most common methods used for controlling the movement of plates during welding. Tacks may be left in and fused with the first run to form part of the completed joint. This means that the tack must be fully penetrated and of adequate length and strength to prevent movement. In the absence of other information tack weld lengths can be two to three times the thickness of the plates being welded.

Jigs are designed mainly for the ease of assembly, but they can also incorporate clamps to prevent the movement of the parts during welding. A good jig will allow good access to the weld area and enable 50% of the welds to be completed before removing the component.

Pre-setting the plates in the opposite direction to the contraction will allow the plates to be pulled back into line as the contraction takes place. It is not easy to estimate the amount of pre-set, but a good guide is the thickness of the plate in the opposite direction to the contraction (see Fig. 1.29).

Figure 1.29 Pre-setting.

Pre-setting of fillet welded joints can be a problem, as it can result in a bad fit-up when the vertical is rocked away from the base plate, leaving a gap. In this case it may be more appropriate to use some form of temporary restraining method, such as clamping, gussets, etc., which can be removed on completion of the weld (see Figs 1.30 and 1.31).

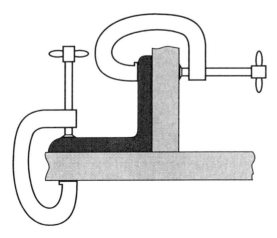

Figure 1.30 Temporary support.

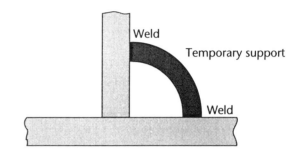

Figure 1.31 Temporary support.

When the job will allow, it may be possible to use pre-setting, in the case of a fillet welded joint, by pre-setting the base plate (see Fig. 1.32). However, this may not always be possible.

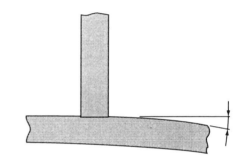

Figure 1.32 Pre-setting of the base plate.

The **back step** method of depositing weld metal to control distortion takes advantage of the principle that as one weld deposit is cooling, the adjacent deposit is expanding. This tends to minimise the effects of distortion (see Fig. 1.33).

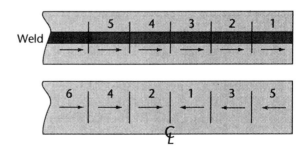

Figure 1.33 'Back step' welding to reduce distortion.

Planned welding sequence is a method used to limit the amount of heat input and, therefore, expansion on the joint in any one place at any one time. Starting from one end of the joint, mark out and number the joint 1,2,3, 1,2,3, 1,2,3 ..., etc., along its length, then starting from the same end commence welding all number ones, then two and so on until the weld is complete (see Fig. 1.34).

Figure 1.34 'Planned welding sequence' to reduce distortion.

Balance weld both sides of the joint where possible, for example, a double 'V' butt, where welds can be deposited alternately each side. 'T'-joints – fillet welded both sides – is also another example. Use as little weld metal as is required to meet design requirements, and use it to the best advantage. Stitch welds may be sufficient in some cases (see Fig. 1.35).

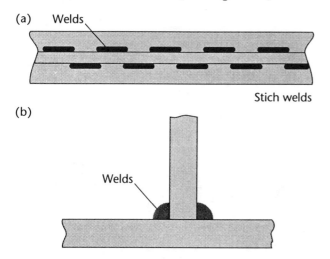

Figure 1.35 (a) Balanced deposition of double 'V' butt weld. (b) Balanced fillet welds on a 'T'-joint by 'stitch welding'.

Back-to-back assembly. Where the shape of the fabrication will allow, it is possible, after tacking up, to use this method of assembly and final welding. As the name 'back-to-back' implies, when the component parts of the fabrication are tacked up, two similar fabrications can be clamped together and welded alternately. The effect of this is again balancing the distortion (stress) on the parts being welded.

Figure 1.36 shows assembled components ready for final welding, balancing the welding,

and therefore the forces, acting on the fabrication. The sequence in which the welds have been numbered for completion also includes the **planned welding sequence** technique.

One or more of these methods can be used to combat the effects of distortion, keep them all in mind, and use them to your advantage.

Do not forget – **prevention is better than cure!**

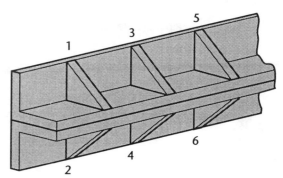

Figure 1.36 Back-to-back assembly.

Weld defects

Defects in welds can result from many different sources ranging from damp or damaged electrodes to soiled joint profiles. This section shows typical weld defects (see Fig. 1.37) and goes on list the possible causes of these defects in Table 1.5.

One possible source of defect to be aware of is the lack of skill of the operator. It is most important to remember that there is little space affordable for operator error when considering the sources of defect which are not directly caused by the operator, but, some of which, the operator may be able to control.

Inspection

Welding inspectors have a wide variety of methods and processes to aid them in the examination of welds. This section is a basic introduction to these processes, but first of all let us consider the two main categories into which all of these methods or processes are covered. These are:

- non-destructive testing (NDT)
- destructive testing (DT).

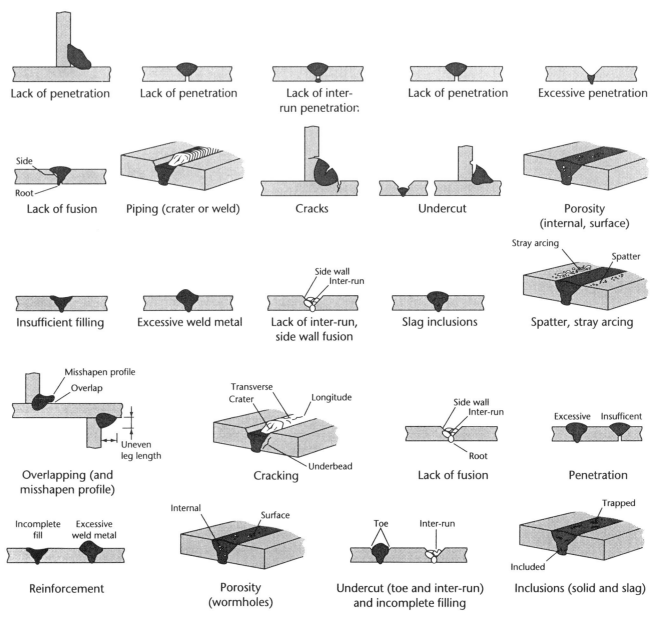

Figure 1.37 Weld defects (labelled as indicated).

Non-destructive testing involves methods where no damage is caused to the weld metal or parent plate leaving the component or structure intact. Some of the methods used include:

a) visual examination
b) magnetic particle examination
c) radiographic examination
d) dye-penetrant examination
e) ultrasonic examination.

Visual examination, although sometimes forgotten, is simply the use of the naked eye to look for visible defects on the surface of the weld. To help determine the extent of any defects and enable improved accuracy a magnifying glass will help improve visibility. Plasticine, when pressed into any surface irregularity, will help to give a more accurate impression of any undercut which may be present. Using the fingernail to scratch the surface will give a good indication of surface smoothness and the evenness of the weld ripple. Visual examination is a widely used

Table 1.5 Typical weld defects

Faults	Causes
Manual metal arc (MMA)	
Undercut	High amperage – oversize electrode – electrode angle – speed – high voltage high amperage – manipulation – incorrect electrode
Slag inclusions	Incorrect cleaning of weld – rusty plate – incorrect electrode angle – manipulation – low amperage – too small electrode (multi-runs) – low voltage
Gas holes, piping	Damp electrodes – damp plate – overheated electrode – wrong type of electrode – rusty plate – manipulation of electrode
Lack of penetration	Incorrect set up of plate – misalignment of plates – too large a root face ('V' joints) – too small a gap ('V' joints) – incorrect technique – too large an electrode – low amperage – low voltage
Lack of side-wall fusion	Technique – electrode manipulation – misalignment of plate – electrode position – electrode angle – small electrode (multi-runs) – slag left in joint
Penetration	Too small an electrode – too fast – manipulation
Lack of interrun fusion	Electrode angle – manipulation of the electrode – electrode position – too small an electrode (multi-runs) – slag left in joint – low amperage low voltage
Lack of reinforcement	Too small an electrode – too fast – manipulation
Cold laps	Low amperage – electrode angle – electrode position
Excess reinforcement	Low amps – speed of travel too slow – electrode too large – incorrect technique
Incorrect profile	Wrong angle of electrode – incorrect weld sequence – incorrect speed of travel – electrode too large
Incomplete fill	Speed of travel too fast – incorrect weld sequence electrode too small (single passes) – not enough passes
Spatter	Amperage too high – too long arc length – damp electrodes – contaminated parent metal
Cracking	Insufficient pre- or post-heat – weld deposit too small – incorrect filler/electrode – restraints on joint–weld metal shrinkage
Metal arc gas shielded (MIG/MAG)	
Undercut	Amperage too high – voltage too high – speed too fast – incorrect torch angle – tip of torch too far from joint – unbalanced arc – technique
Gas holes/piping	No shielding gas – gas pressure too low – gas pressure too high – part of torch tip broken – oil/water on joint – rust in joint – rusty filler wire – incorrect torch angle – incompatible wire/plate – gas shield blown away – silicon left in joint (steel) – technique
Lack of penetration	Too small a gap – too large a root face ('V' joints) – plate misalignment – filler wire too large – low amperage/voltage – incorrect plate preparation – technique
Excess penetration (wire burn through)	Too large a gap – too narrow a root face – high amperage/voltage – unbalanced wire feed speed – speed of torch movement too fast/slow – too small a filler wire – technique
Lack of side-wall fusion	Incorrect torch angle – manipulation of the torch – incorrect joint preparation – torch out of position – dirty joint – silicon left in joint (steel) – a low voltage – technique
Lack of inter-run fusion	Incorrect torch angle – low voltage – low amperage/wire speed – silicon left in joint (steel) – torch out of position with joint – manipulation of torch – technique
Lap	Low amperage/voltage – incorrect torch angle – incorrect torch position – dirt in joint – technique

Table 1.5   Continued

Excess reinforcement	Incorrect balance of volts and amps – incorrect technique
Incorrect profile	Wrong torch angle – incorrect speed of travel – incorrect weld sequence
Incomplete fill	See MMA
Spatter	Incorrect volts/amps setting, common on spray transfer
Cracking	See MMA
Slag inclusions	Small amounts of silica slag but no real problem
Irregular stop/start	Poor technique – inadequate instruction and training
Tungsten gas shielded (TIG)	
Undercut	Torch angle – high amperage – manipulation of torch and filler rod – too small a filler rod speed – too fast – incompatible filler rod and metal
Gas holes	No shielding gas – dirty plate incompatible filler rod and metal – technique – manipulation of torch – speed too fast – shielding gas low – gas shield being blown away – no gas shield on root ('V' joints)
Excess penetration	Gap too large – speed slow – too large a tungsten electrode – too small a root face ('V' joints) – amperage too high – torch and filler rod manipulation
Lack of interrun fusion	Torch and filler rod manipulation – incorrect torch angle – torch and filler rod position – low amperage – too large a filler rod
Poor or no penetration	Gap too small ('V' joints) – too large a root face ('V' joints) – misalignment of plates – low amperage – low voltage – torch and filler rod manipulation – incorrect joint preparation, i.e. too narrow a 'V' preparation ('V' joints)
Lack of side-wall fusion	Technique – torch and filler rod manipulation – torch angle – torch position – low amperage
Excess reinforcement	Low amperage – excess filler
Incorrect profile incorrect weld sequence	Wrong torch angle – excess filler – incorrect speed of travel –
Incomplete fill	Speed of travel too fast – incorrect weld sequence – not enough filler (single passes) – not enough passes
Cracking	See MMA
Slag inclusions	Not possible (may be solid inclusions)
Oxyacetylene (OA)	
Undercut	Overheated plate – excessive flame temperature – oversize nozzle – torch angle – speed too fast – incorrect flame – gap under vertical plate (fillets) – technique
Gas holes/piping	Incorrect flame – rusty plate – oil on plate – incompatible filler rod/plate – water on plate – speed of movement too slow/fast
Lack of penetration	Plate misalignment – too small a gap/root face ('V' joints) – too small a nozzle – too large a filler rod – incorrect flame – low heat in flame – torch angle – technique
Lack of side-wall fusion	Torch angle – position of torch – low flame heat – incorrect flame – incompatible filler rod/plate – oil/dirt on plate – plate preparation – technique
Excess penetration	Too much filler – incorrect torch angle – work too cold
Incorrect profile sequence, torch angle	Incorrect: angle of filler wire, welding speed, welding
Incomplete fill	See TIG
Cracking	See MMA

method of NDT and is probably used without thinking. The limitations are that it is only suitable for detecting surface defects (Fig. 1.38).

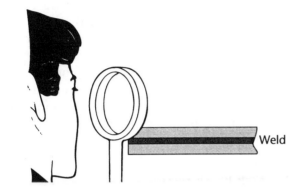

Figure 1.38 Visual examination.

Magnetic particle examination, uses a magnet to magnetise the area to be examined, e.g. the weld area. Magnetic particles suspended in a solution from an aerosol can or container are applied to the area to be examined. Any cracks in the area between the two magnetic poles will cause a change in polarity at the crack resulting in a collection of particles at this point. Cracks are easier to detect when they are at right angles to the lines of the magnetic field. For this reason it is good practice to change the position of the magnet as progression is made along the joint. This process is limited to metals capable of being magnetised and for the detection of surface defects only (see Fig. 1.39).

Radiographic examination, uses a source of radiation supplied either by an electrical power supply or a radioactive material. The rays produced have a high energy value capable of penetrating metal and projecting an image to a light-sensitive film. If a defect is present in the weld, some of the light will be absorbed allowing less to pass through to the film to show a darker image. Where the metal is solid with no defect, a lighter image is projected. Interpretation of these light and dark images demands great skill and experience on the inspector's part. This method is mostly used for detecting internal defects. Strict safety rules and regulations must be followed when using radiographic inspection techniques. Where the radiation is supplied by an electrical power supply the rays are referred to as X-rays. Where the radiation is supplied by a radioactive material the rays are referred to as gamma rays (see Fig. 1.40).

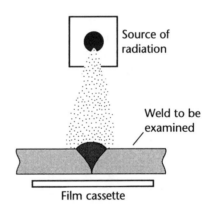

Figure 1.40 Radiographic examination.

Dye-penetrant examination, in its simplest form, uses three aerosol cans containing:

i)   cleaner
ii) developer (usually white)
iii) dye or penetrant (usually red).

The surface to be examined is thoroughly cleaned, allowed to dry and a coat of dye is applied to the suspect area. This is allowed to stand for 15–20 minutes to allow penetration of any defects that may be present. The surface is then cleaned, dried and sprayed with a thin layer of developer, and any dye absorbed by a crack will be drawn up into the developer giving an indication of the crack location. Thick

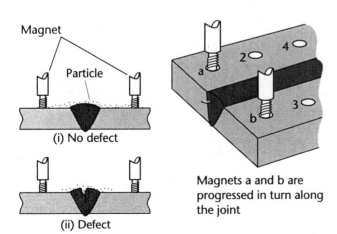

Figure 1.39 Examination using magnetic particles.

layers of developer will distort the image and give an incorrect interpretation of the defect (see Fig. 1.41).

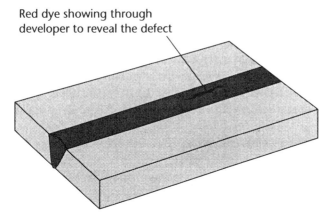

Red dye showing through developer to reveal the defect

Figure 1.41   Dye-penetrant examination.

Ultrasonic examination, uses sound waves to penetrate the weld area. These ultrasonic waves are capable of being transmitted through fairly thick sections but are deflected by the presence of a defect in their path. The method uses a transmitter and a receiver connected to a cathode ray oscilloscope which will give a visual display of the presence of the defect (see Fig. 1.42).

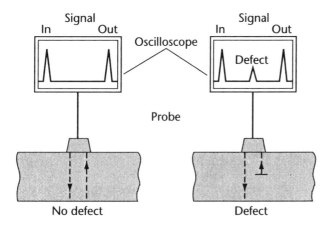

Figure 1.42   Ultrasonic examination.

The area being examined has to be relatively clean and flat. This may involve using some form of surface preparation. A thin coating of oil is sometimes used to aid the transmission and receiving of the ultrasonic waves. As the

waves pass through the metal with no interruption a steady horizontal line appears on the visual display. If the waves are interrupted by a defect, the horizontal line on the display will show a peak in its form, indicating the presence of the defect. The skill of a trained inspector is needed to establish a correct interpretation of the visual display.

Destructive testing can only be carried out by removing samples from the welded joint or an extension of it. Samples or coupons have to be cut out to perform the required tests. These include:

a) bend tests
b) nick-break tests
c) macro-examination
d) tensile tests (weld metal and parent metal)
e) impact tests.

Bend testing will involve cutting samples from the joint which include both weld metal and parent metal. Bend testing is a collective name which includes root bends, face bends, side bends, longitudinal bends (along the weld) and transverse bends (across the weld). Sample dimensions are usually listed in the code or standard being used, e.g. BS4872. Any defects present will show up as depressions or openings on the surface in tension, openings from the edges or across the face where lack of fusion has occurred (see Fig. 1.43).

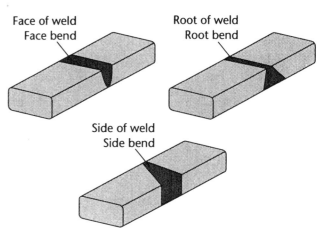

Figure 1.43   Bend test.

Nick-break tests usually involve cutting one or more samples from the weld. These samples will include a stop and start of the weld. The surface of the weld is grooved to locate the point of fracture. It is then broken open to reveal the internal or fracture faces. The type of defect revealed will be lack of fusion, slag inclusions, porosity.

Macro-examination is carried out by removing a sample cross-section of the welded joint. This is then polished by filing and various grades of emery cloth to achieve a smooth, scratch-free surface. When this has been achieved do not touch the surface as this will affect the quality of the image. At this stage an etching solution is used to coat the surface of the sample to reveal the contour of the weld. Any large defects will be seen by the naked eye; improved vision can be achieved by the use of a magnifying glass (or microscopic examination). This will show up the granular structure of the weld and parent metal. Other information obtained from a macro-examination will include the heat-affected zone and number of weld deposits (see Fig. 1.44).

strength of the deposited metal when partially diluted by the parent metal. In both cases the samples are usually tested to destruction and in the latter of the two test situations the fracture will take place in the parent metal (see Fig. 1.45).

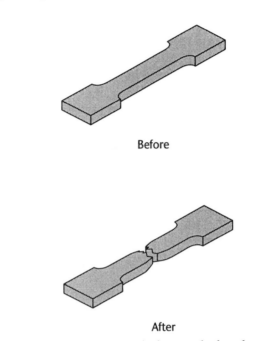

Before

After

Figure 1.45 Tensile test before and after fracture.

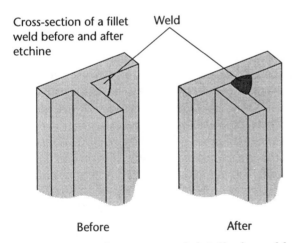

Cross-section of a fillet weld before and after etchine

Weld

Before   After

Figure 1.44 Etching to reveal detail of a weld.

Tensile tests basically consist of two types, these are weld and parent metal test and all-weld metal test. The all-weld metal test pieces tend to be small and involve some quite exacting machining of the sample and accurate recording of the test results to determine the strength of the filler metal after deposition. The weld and parent metal test examines the whole of the weld area to determine the

Impact tests involve the fracture of a previously notched test piece. The most common form of this test is the 'Charpy impact' test. A specimen is removed from the welded joint and machined to specific dimensions. A notch is then machined in the specimen to locate the fracture to the precise point on the joint for which the information is required, for example, weld centre line, heat-affected zone, weld junction, etc. The purpose of the test is to determine the brittleness or toughness of the weld and parent metal, and can be carried out at a range of temperatures from −20 to +20°C to determine the material behaviour at specific temperatures. The test piece is placed in a test machine supported by the anvil. This is then struck by a hammer released from a predetermined height. The hammer is connected to a dial indicator which will measure the amount of energy absorbed in breaking the test piece. At −20°C little energy will be absorbed in breaking the test piece, the metal is said to be brittle at low temperature (see Fig. 1.46).

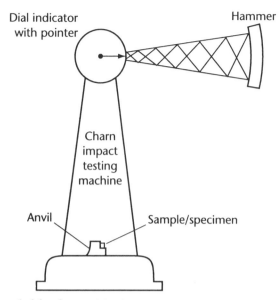

Figure 1.46   Impact test.

Welding procedure and welder qualifications

When designing a fabricated structure, one of the principal factors to consider is the service conditions, i.e. what are the conditions in which the fabrication has to function. Such conditions may range from extreme high to extreme low temperatures, or both. Whatever the circumstances, designers, engineers, fabricators, welders and, not least of all, the customer, must be confident that the finished welded structure has the desired properties to withstand the service conditions. This is achieved by choosing the correct raw materials, in this case steel, selecting the most suitable fabricating processes/techniques and working procedures, all of which contribute to achieving a product of the required properties and quality.

Achieving the properties of the fabrication depends to a large extent on the way the welds are made. Production of welds at this level is controlled by the use of tried and tested welding procedures and qualified welders.

Everyone involved in the production process will be affected by working procedures. Welding procedures are particularly important to ensure weld quality and will specify everything from material specification through to inspection standards.

The person most obviously required to com-

ply with the welding procedure is the welder, as this will set out the previously tried and tested methods required to achieve the desired quality.

Some of the requirements listed on the procedure that will directly affect the welder are:

a) Welding processes: ensure that the correct process is being used.
b) Joint type: does this comply with that specified on the procedure?
c) Weld run: sequence/position of weld runs in the preparation.
d) Metal thickness: as stated on the procedure.
e) Filler metal type: ensure compliance with procedure.
f) Filler metal classification: as procedure.
g) Gas/flux type (if used): as specified.
h) Welding current AC–DC: as specified.
i) Welding parameters: amps and volts.
j) Welding polarity.
k) Welding position.
l) Pre-heat/post-heat and interpass temperatures.

The above list is by no means exhaustive, all of these may well be considered as the welder's responsibility and are considered essential criteria in the production of quality welds.

The tinted areas shown in Table 1.6 are details of which the welder should be aware, although they may not all be effective at the same time. For example, when MMA welding gas shielding will not be relevant.

Pre-heating may be the responsibility of other departments or even companies specialising in these methods. On the other hand, observing the stated pre-heat, interpass and post-heat temperatures may well affect the successful achievement of acceptable welding conditions, resulting weld quality and service conditions of the component.

The prime function of a weld procedure is to achieve a level of quality applied to the production of a welded component or structure.

The prime function of a welder qualification is to approve a welder to carry out the welds on a welded structure to a set quality standard.

A weld procedure in its simplest form is a list of requirements laid down that produce a welded joint achieving the desired mechanical properties required to maintain a 'safe in oper-

Table 1.6 Welding procedure approval – test
certificate

Manufacturer's welding procedure reference No.:
Examiner or test body
Reference No.:
Manufacturer:
Address:
Code/testing standard:
Date of welding:
Extent of approval
Welding process:
Joint type:
Parent metal(s)   Conditions of tempered:
Metal thickness (mm):
Outside diameter (mm):
Filler metal type:
Shielding gas/flux:
Type of welding current:
Welding positions:
Preheat:
Post-weld heat treatment and/or ageing:
Other information:

Certified that test welds prepared, welded and tested satisfactorily in accordance with the requirements of the code/testing standard indicated
above.

Location	Date of issue	Examiner or test body Name, date and signature

---

ation' lifespan of a component or structure to
which the weld is applied.

The list of requirements will specify some or
all of the following:

■ Code or test standard being applied.
■ Range of approval.
■ Welding process.
■ Joint type and welding sequence.
■ Parent metal group.
■ Parent metal thickness.
■ Filler metal type and designation.
■ Gas/flux type.
■ Type of welding current.
■ Welding current/volts.
■ Welding position.
■ Pre-heat/post-heat temperatures.

Other supplementary information may be
required to satisfy specific needs.

In order to carry out a manual weld procedure test the first thing needed is a welder. The
person chosen to carry out the weld test is usually someone with previous experience and in

some cases previously qualified to other procedures. This is to ensure that the procedure laid
down is followed correctly in order to achieve
the required outcomes. If the outcome and
results of NDT and DT examination are successful, it is possible to approve the procedure
and the welder in one operation.

Welding procedure and welder qualification
tests are usually carried out to national or international standards. Some of the more common
or most widely used standards are the British,
US and European standards which include:

■ BS4872 – Approval of welders when welding
procedure approval is not required.
■ BS5135 – Arc welding or carbon and carbon
manganese steels.
■ BS2971 – Class II arc welding of carbon steel
pipework for carrying fluids.
■ ASME Boiler and pressure vessel code,
Section IX. Welding and brazing qualifica-
tions.
■ BSEN 287 – Approval and testing of welders
for fusion welding.
■ BSEN 288 – Specification and approval of
welding procedures for metallic materials.

These are a sample of the numerous standards you may encounter, or be tested to, in
your career as a welder.

Tests used in approving a weld procedure have
been explained earlier in the section on Inspection (page 19) and include the following:

■ Visual examination   NDT.
■ Radiographic examination NDT.
■ Bend testing   DT.
■ Macro-examination   DT.
■ Tensile tests   DT.
■ Impact tests   DT.

Other tests may include nick-break and hard-
ness tests depending on the joint type or standard being used.

The tests involved in welder approval to a
procedure already qualified would include:

■ Visual examination.
■ Radiographic (depending on level of
approval).
■ Bend tests.
■ Macro-tests.
■ Nick-break (depending on level and joint
type).

Welders may experience many qualification tests during their career. It is not like an academic examination that once you have passed it lasts for ever. If you change jobs, process metals, consumables, joint type, thickness of metal, etc., you may find yourself doing another welder approval test to a different procedure.

The purpose of the welder qualification test is to measure the welder's ability and competence to produce welds to a specific quality or procedure, thus providing a level of welding quality control applied to a component or structure.

Weld quality is of utmost importance and at the forefront of all design engineers' criteria. It is also important to the welder. One of the 'in' phrases when I was on the welding circuit was that 'you are only as good as your last butt'.

2

Manual metal arc welding

Manual metal arc welding equipment (see Fig. 2.1)

Manual metal arc (MMA) welding uses both alternating current (AC) and direct current (DC). Generally AC is associated with shop work and DC with site work, as the DC supply in the form of a generator is much more portable. Also with the lower open circuit voltage (OCV) of DC it is safer in damp, confined or dangerous conditions, where there is a higher risk of electric shock or accidents resulting from electric shock. AC current will tend to pull the user into the electrical supply, DC will push the user away.

A DC supply gives the operator a choice of polarity: electrode positive (+ve) or electrode negative (−ve). This means that the electrode can be connected to the +ve or −ve output terminal of the power source.

Connecting the electrode to the +ve terminal will result in two-thirds of the arc heat being concentrated at the electrode and one-third at the work. Connecting the electrode to the −ve terminal will result in one-third of the arc heat being concentrated at the electrode and two-thirds at the work. A typical use for the latter is for the first or root runs in open butts with low-hydrogen (hydrogen-controlled) electrodes on plates and/or pipes.

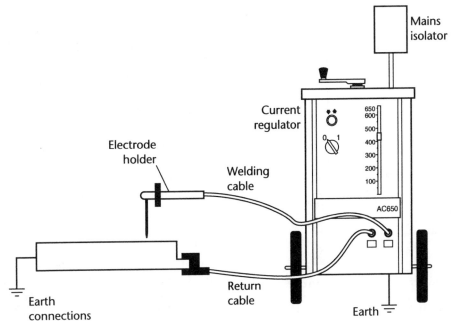

Figure 2.1 Manual metal arc welding equipment.

One of the disadvantages of using a DC supply is the occurrence of arc blow. This is when metal is expelled from the arc or molten pool caused by magnetic fields set up in the work. The effect of this can be reduced by:

a) changing the direction of welding
b) moving the position of the return clamp
c) using smaller electrodes
d) wrapping the return cable around the work or return clamp.

Mains electricity as supplied is not suitable for welding purposes, as it is supplied at high voltage and low amperage. For welding the reverse is required. Therefore, the purpose of the welding plant is to reduce the voltage and increase the amperage.

There are three main types of welding plant:

■ transformer = AC input to suitable AC output
■ rectifier = AC input to suitable AC and DC output
■ generator = generated DC only.

Over this range of equipment the open circuit voltage (OCV) ranges from 40 to 100 volts, with an arc voltage of 25 to 40 volts – which is the voltage required to maintain the arc.

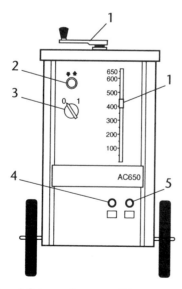

Figure 2.2 AC transformer: (1) current selector control handle; (2) remote switch socket; (3) on/off switch; (4) return cable socket; (5) electrode cable socket; (6) current indicator.

Transformer

The purpose of a transformer is to provide a single function of changing the high voltage/low amperage mains supply to one of low voltage/high amperage welding supply. The OCV will vary depending on the size and type of equipment, but usually ranges from 40 to 100 volts. This is the voltage available for the initial striking of the arc. Once the arc is established the voltage will drop to between 25 and 40 volts. This remains constant while the amperage will have adjustments of between 45 and 110 amps on smaller sets, to 50–350 amps and over on larger sets.

Most modern welding sets have a total variable amperage control; some of the older types, however, may have stepped-type adjustments or plug-type settings.

Rectifier

The rectifier can be a separate piece of equipment connected to a transformer, although in modern welding plant the transformer and the rectifier are usually contained in one package referred to as a transformer/rectifier. The purpose of the rectifier is to change the previously transformed AC mains supply into DC for welding.

Generators

Generators are usually driven by a primary source such as an electric motor from the mains supply, or engine driven, the latter being the most popular, as this gives much more flexibility to the welding set for site operations where a mains supply is not always available. The OCV generated is between 40 and 60 volts, and is normally adjustable between these limits.

Current capacity will vary with the size of the generator, 150–400 amp range generators are common.

A piece of equipment which combines the transformer and the rectifier and is known as the **transformer/rectifier** is quite popular for static situations. This offers the welder a choice of suitable welding current in both AC and DC supply. This equipment also allows the flexibility of AC and DC with less moving parts than the generator, thus reducing maintenance costs. With correct maintenance the transformer/rectifier will give many years of service.

Without placing themselves in dangerous situations, welders should always check their plant is in satisfactory working order. Any damages to cables, switches, insulation, any power loss, overheating, noise or excessive vibration should be reported to the supervisor.

Open circuit voltage (OCV)

The maximum OCV allowed for MMA welding is 100 volts, normally associated with the AC power supply. Slightly lower OCVs are used with the DC power output:

■ AC 50–80 OCV
        80–100 OCV
■ DC 60–80 OCV

The MMA process, or stick welding process as it is commonly known, is well established and widely used. The equipment is relatively simple, making the process very versatile and mobile, suitable for shop and site work. The process uses a covered electrode to create an arc between the end of an electrode and the plates to be joined. The arc is the source of heat which melts the edges of the plates. The whole of this molten area is known as the weld pool. The covered electrode is used up during this process, melting the electrode material and plate edges to form the weld.

With all welding processes, the base metal, the weld metal being transferred across the arc to the weld pool and the end of the electrode are all exposed to contamination from the atmosphere which will allow impurities to combine with the molten metal. This results in reducing the mechanical properties (strength) and corrosion resistance of the weld, and possibly of the base metal. This means that the molten weld area, the end of the filler or electrode and the transferring metal must be protected from atmospheric contamination. In the four welding processes being considered, this protection is achieved in different ways.

In MMA welding the arc is shielded by gasses produced by the burning of the electrode coating (see Fig. 2.3). These gasses com-

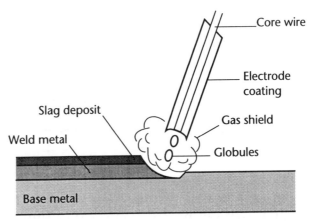

Figure 2.3 Principle of manual metal arc welding.

pletely cover the end of the electrode and the weld area to form a protective environment in which the weld is deposited. Other functions of the electrode coating are to act as a fluxing agent to clean the weld area, and for carrying additional elements to the weld. These additional elements can exist as iron powder particles which will increase the amount of metal deposited for a given size of electrode (increase its metal recovery), to improve mechanical properties or corrosive resistance. The end result of the melted coating is slag forming on the surface of the deposited weld which then solidifies on cooling. The slag will protect the metal during cooling and, if not removed, will give some degree of protection to the finished weld until a surface protection is applied. The coatings on electrodes used for positional welding have what is known as a 'quick freezing slag' so that, when welding in the vertical position, for example, the slag will give some help in forming the shape of the weld deposit. However, it must be recognised that this does not reduce the level of operator skill required when positional welding.

Striking the arc

For the beginner, initial striking of the arc can be a strange experience from the point of view that with the common type of welding screen it is not possible to see the position of the end of the electrode in relation to the joint until the arc is struck. The following simple rules will help overcome this initial problem (see Fig. 2.4).

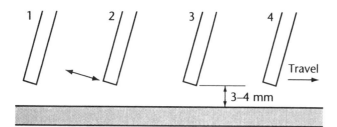

Figure 2.4   Striking the arc.

Locate the end of the electrode over the point where the weld is to start (position 1). Now bring the welding screen in front of your face to protect your face and eyes. Lower the electrode onto the plate with a gentle stroking motion (position 2). The arc will then be established.

Lift the electrode until the visible arc length is 3–4 mm (position 3). The arc should then be stabilised (kept burning) at this level before moving forward in the direction of travel (position 4). It is important to maintain this distance between the end of the electrode and the work. This is known as the arc length.

When the electrode has been consumed (used up), a new electrode is placed in the electrode holder and the same procedure is followed as in (1) to (4) above. The start position of the new weld (position 1) is slightly in advance of the previous weld run. Once the arc is established, move quickly into the weld pool of the previous weld and move along the joint. This point on the weld is called the stop/start point. By adopting this method at stops and starts it will enable continuity of the weld size and help reduce the effect of weld defects at this point (see Fig. 2.5).

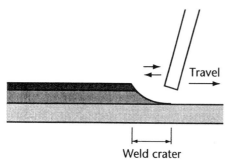

Figure 2.5   Starting a weld.

Stop/starts

Stop/starts should be as smooth as possible, although a slight bump will be seen. To keep this to a minimum it is important that movement into the crater of the previous weld is done as quickly as possible once the arc is stabilised (see Fig. 2.6).

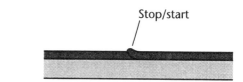

Figure 2.6   Stop/starts.

Electrode angles

When making a weld, either on a flat plate or between two pieces of metal, one of the basic, but most essential, skills of which to gain mastery is electrode angles.

There are two angles which affect the successful deposition of the weld metal. These are the angles to either side of the electrode, known as the **tilt** angle and the angle between the electrode and the plate surface in the direction of travel, known as the **slope** angle (see Fig. 2.7).

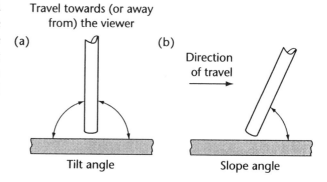

Figure 2.7   (a) The tilt angle. (b) The slope angle.

Although these angles will be affected by many other things in the welding process, they will play an important role in the successful deposition of the weld bead, regardless of the type of process being used. It is therefore important to learn the skill of electrode angle control at an early stage in the development of welding competence.

There is a base point from which this skill can be developed, and Fig. 2.8 shows the electrode angles suitable for depositing the first or root deposit in a fillet welded 'T' joint.

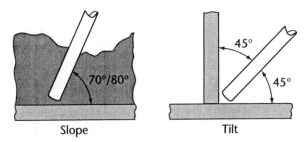

Figure 2.8 First deposit in a fillet welded 'T' joint.

When welding with covered electrodes, the electrode coating melts off in the arc to re-form and solidify on the weld surface in the form of a slag, helping to protect the weld metal from contamination by the atmosphere during cooling and, to some degree, the slag will protect the weld metal after solidification has taken place. It is only a temporary protection; permanent surface protection coverage should be considered in the long term.

Controlling the flow of this molten slag is achieved by the 'slope' angle of the electrode. If the slope becomes too steep there is a tendency for the molten slag to run around and in front of the electrode and into the weld pool (see Fig. 2.9). This will tend to form a barrier between the end of the electrode and the plates being welded, preventing penetration into the plate surface or root of the joint. It is also likely to result in weld defects, such as slag inclusions, lack of penetration and lack of fusion.

The effect is common to any of the welding processes where there is some form of flux medium for protecting the weld metal during deposition, in MIG/MAG when using flux-cored wires, oxyacetylene welding where fluxes are used or MMA welding with covered electrodes.

When depositing a weld bead onto a plate surface, or along the joint line between two plates, it is important that the molten weld metal is deposited equally to either side of the joint. In order to get the required penetration into each plate, imagine the electrode as a pipe, and the molten metal as water flowing from it. The water will flow whichever way you direct the pipe; the same effect happens with the electrode and the molten metal. The means of controlling this is by the control of the 'tilt' angle.

If the electrode is moved to either side of the centre line of the joint, the molten metal will tend to do the same. This will have the effect of the weld metal being deposited to one side of the joint line. This results in missing the plate edges and causing weakness in the joint area (see Fig. 2.10).

(a)

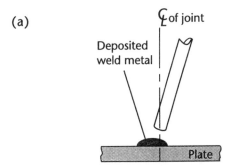

(b)

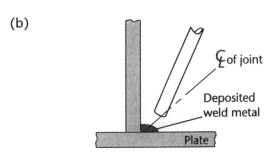

Figure 2.10 (a) The effect of tilt angle on butt joints. (b) The effect of tilt angle on 'T' joints.

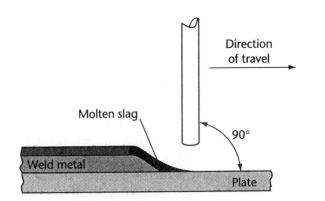

Figure 2.9 The effect of slope angle.

Some of the defects associated with this are missed plate edges, lack of side wall fusion and unequal leg length. All these will have a serious effect on the strength of the joint and its performance in service.

This section has considered the electrode angles of 'slope' and 'tilt' and the problems associated with poor control of these angles. It is also important to remember that controlling the 'tilt' angle can be used to our advantage, especially in the case of multi-run welds. By varying the 'tilt' angle slightly, it is possible to direct the molten metal to the desired area of the weld joint.

In Fig. 2.11 *A* shows the tilt angle for deposition of the first or root run in a 'T' joint. *B* shows the 'tilt' angle for depositing the second run in a multi-run fillet welded 'T' joint.

Slope and tilt angles can work for, and against, the welder. Gaining control and being aware of the effects in changing 'slope' and 'tilt' angles will be most rewarding and beneficial to the skilled welder in controlling the weld pool and achieving sound quality welds of correct shape and size.

Weaving

Weaving of the electrode refers to the side-to-side movements of the electrode. This is sometimes required to spread the molten metal across the joint width or, in the case of position welding, to control the heat input.

Textbooks show many different weave patterns which may be suitable in various situations, and some of these are shown in Fig. 2.12.

The maximum weave width for a given size of electrode is two to three times its diameter. Any greater than this and the edges of the joint tend to become cold. This gives a coarse weave pattern or, even worse, lack of fusion, fusion penetration or inter-run fusion.

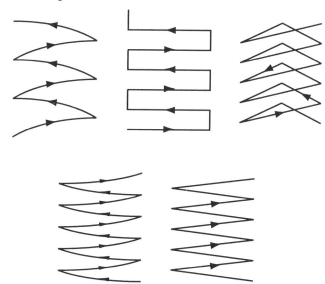

Figure 2.12   Various weaving patterns.

When a weave pattern is used to suit joint requirements, this pattern is usually reflected in the finished weld. If the movements are erratic or inconsistent then the finished weld appearance may suffer from what could be called a weld defect. One example of this is shown in Fig. 2.13.

If the progression along the joint is progressed in a coarse movement, the weld appearance will be as shown in Fig. 2.13.

Steady progression along the joint will achieve a smoother weld profile and a complete filling of the joint (see Fig. 2.14).

Slight pauses at the sides of the weave will also help in this situation, but if this is excessive, overheating may occur and the molten metal may sag or overlap the edges of the

(a)

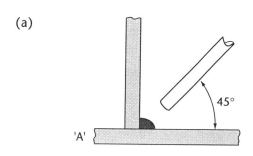

(b)

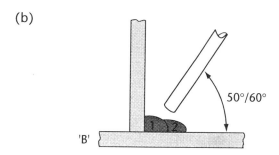

Figure 2.11   (a) A multi-run weld; first pass. (b) A multi-run weld; second pass.

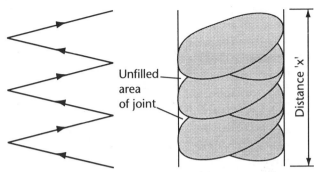

Figure 2.13 The undesirable effects of coarse weaving.

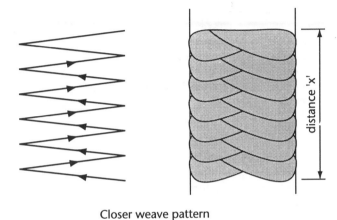

Closer weave pattern

Figure 2.14 Superior joint obtained by fine weaving.

joint. Try to keep weaving to a minimum where possible as when welding the size of the molten pool will extend slightly beyond the diameter of the electrode (see Fig. 2.15).

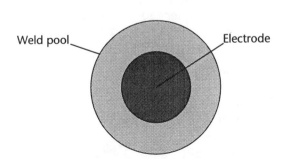

Figure 2.15 Weld pool is always larger than the electrode itself.

Electrode classification

Most welders will know a particular electrode and its performance capabilities (amperage, position, polarity, etc.) by its trade name. This familiarity gained through experience and use.

Although this may be the case, it is not totally acceptable in all situations, and for this reason a more standardised method of recognition is needed.

All MMA electrodes are covered by a classification scheme, according to their flux coatings, welding position suitability, current capacity and mechanical properties. The specifications for MMA welding electrodes are covered in BS639 (1986) Electrodes for MMA welding of Carbon and Carbon Manganese Steels. An example of a typical electrode specification is given on page 36.

Electrode size and current capacity

Something else important to the welder about the stick welding electrode is the amperage range for a particular size of electrode. This information is usually listed on the side of the electrode carton and, therefore, readily available. But with experience the welder will get to know the amperage ranges. The size of the electrode refers to the diameter of the core wire (see Fig. 2.16).

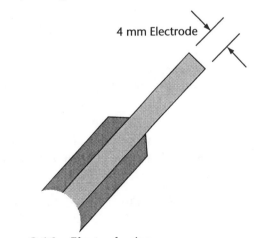

4 mm Electrode

Figure 2.16 Electrode size.

Typical current ranges for MMA electrodes are:

Size (mm)	Current (amps)
2.0	40–65
2.5	70–95
3.25	90–130
4.0	140–180
5.0	160–240
6.0	220–300

BS639

Note: All weld metal mechanical properties.
The mechanical properties of the deposited weld metal refer to ALL-WELD METAL PROPERTIES when deposited in the flat position. These may have little relevance to the properties of a real joint achieved in practice since this will depend on the dilution from the base material, welding position, bead sequence and heat input. Apart from their use for quality control purposes, the mechanical properties of the all weld metal test provide the designer with an initial guide to the selection of electrodes.
This is particularly important in regard to Charpy impact grading. Thus electrodes which have the highest gradings are more likely to offer better Charpy properties when used in practice. They will not necessarily give the Charpy results in a joint which they give in an all weld metal test piece.
BS639:1986
This specification is divided into two parts, the compulsory strength, toughness and covering (STC) code which appears on each electrode and the additional coding showing efficiency, welding positions, power supply requirements and when appropriate, hydrogen control. Use of this classification system is best illustrated by an example:

STC code	Additional	
Ferromax	**E5144BB**	[130 3 1 H]

Consumable Type – **E**
E is for covered electrode for manual metal arc welding. No other consumables are covered by this specification.

Strength – **51**

Designation	TS N/mm^2	min YS N/mm^2	min elongation, % when 3rd digit in classification is		
			0.1	2	3, 4, 5
E43xxx	430–550	330	20	22	24
E51xxx	510–650	360	18	18	20

Third Digit – Impact Value – **4**

Digit	Temp. C, for 28J ave CVN.
Exx0xx	not specified
Exx1xx	+20
Exx2xx	0
Exx3xx	−20
Exx4xx	−30
Exx5xx	−40

Fourth Digit – Impact Value – **4**

Digit	Temp. C, for 47J ave CVN.
Exxx0x	not specified
Exxx1x	+20
Exxx2x	0
Exxx3x	−20
Exxx4x	−30
Exxx5x	−40
Exxx6x	−50
Exxx7x	−60
Exxx8x	−70

Covering – **BB**

ExxxxB	Basic
ExxxxBB	Basic, high recovery
ExxxxC	Cellulosic
ExxxxR	Rutile
ExxxxRR	Rutile, heavy coated
ExxxxS	other types

Efficiency – **130**
The normal electrode efficiency is the ratio of the mass of weld metal to the mass of core wire consumed for a given electrode. It is quoted to the nearest 10% and is included if the figure is equal to or greater than 110.

Welding position – **3**
1 all positions
2 all positions except vertically down
3 flat and horizontal–vertical
4 flat
5 flat, vertically down and horizontal–vertical
6 any position or combination not included above

Power supply requirement – **1**

Digit	Polarity for DC	OCV for AC
0	as recommended	not suitable
1	+ or −	50
2	−	50
3	+	50
4	+ or −	70
5	−	70
6	+	70
7	+ or −	80
8	−	80
9	+	80

Hydrogen controlled electrodes – **H**
The letter H is included for electrodes which deposit not more than 15 ml diffusible H$_2$/100 g deposited weld metal.
Courtesy of BSI, Milton Keynes.

Electrode coatings

Electrode coatings for MMA electrodes are grouped into four main types. These are cellulosic, rutile, basic and iron powder.

Cellulosic electrode

Cellulosic electrode coatings are made of materials containing cellulose, such as wood pulp and flour. The coating on these electrodes is very thin and easily removed from deposited welds. The coating produces high levels of hydrogen and is therefore not suitable for high-strength steels. This type of electrode is usually used on DC+ and suited to vertical down welding.

Rutile electrodes

Rutile electrodes, or general-purpose as they are sometimes known, contain or have coatings based on titanium dioxide. These electrodes are widely used in the fabrication industry as they produce acceptable weld shape and the slag on deposited welds is easily removed. Strength of deposited welds is acceptable for most low-carbon steels and the majority of the electrodes in this group are suitable for use in all positions.

Basic or hydrogen-controlled electrodes

Basic or hydrogen-controlled electrode coatings are based on calcium fluoride or calcium carbonate. This type of electrode is suitable for welding high-strength steels and the coatings have to be dried. This drying is achieved by baking at 450°C holding at 300°C and storing at 150°C until the time of use. By maintaining these conditions it is possible to achieve high-strength weld deposits on carbon, carbon manganese and low-alloyed steels. Most electrodes in this group deposit welds with easily removable slags, producing acceptable weld shape in all positions. Fumes given off by this electrode are greater than with other types of electrodes.

Iron powder electrodes

Iron powder electrodes get their name from the addition of iron powders to the coating which tend to increase efficiency of the electrode. For example, if the electrode efficiency is 120%, 100% is obtained from the core wire and 20% from the coating. Deposited welds are very smooth with an easily removable slag; welding positions are limited to horizontal, vertical fillet welds and flat or gravity position fillet and butt welds.

Irrespective of the type of coating, electrodes need to be handled and stored with care. If they are subject to knocks or mishandling then the coating will be damaged or chipped, possibly making the electrode unusable.

Electrode coatings can sometimes be off centre, in this case the electrode coating will tend to burn away faster on one side during use. This will cause the arc to be uncontrollable causing possible defects in the deposited weld (see Fig. 2.17).

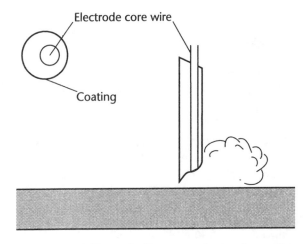

Figure 2.17   Effect of off-centre electrode coatings.

Welding positions for plates

Several examples of welding positions for plates are given in Fig. 2.18 overleaf.

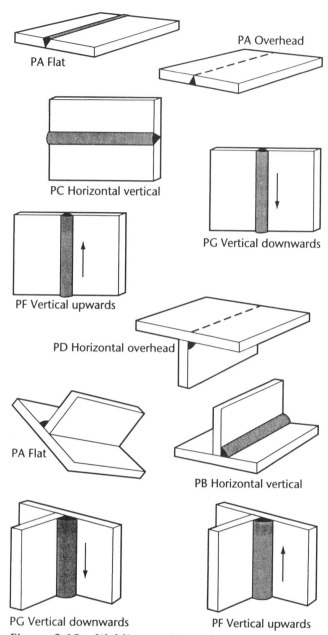

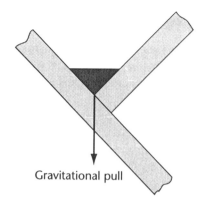

Figure 2.19 Gravity can be used to assist deposition of the molten metal.

Figure 2.18 Welding positions for plates.

operation, it enables increased welding speeds, higher current, larger weld metal deposits per run with the use of larger diameter electrodes, and reduces operator fatigue. The size of the component or fabrication will determine whether or not work manipulation is possible or economical.

Where very large or fixed fabrications are being welded, manipulation will not be possible and welding has to be carried out in position. The MMA process in the hands of a skilled operator lends itself very favourably to welding in all positions.

Plate positions

When welding in the various positions the slope angle of the electrode is usually 60–80° to the surface of the plate, in the direction of travel. The tilt angle to either side of the electrode will vary slightly, depending on the type of joint and positioning of the weld deposit in the joint. Figure 2.20 gives some examples of electrode angles for the various welding positions.

Positional welding

With all of the welding processes considered, where possible, the work should be positioned so that the weld can be made in the flat or gravity position. This ensures that as the molten weld metal is deposited, it is held in position by gravity (see Fig. 2.19). This may mean that the component or fabrication has to be rotated or manipulated into a suitable position. Although this involves an additional

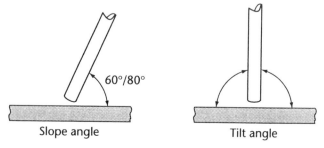

Figure 2.20 Usual electrode angles for plate welding.

Flat position

Horizontal–vertical 'T' fillet joints (10 mm) welded in the flat position require a slope angle of 60–80°, and an equal tilt angle between each side of the electrode and the plate surfaces for the first or root run. No weaving is required on this deposit and only the tilt angle will require variation to overcome any tendency the molten metal may have to fall toward the horizontal plate surface causing overlap. The tilt angle should be maintained along the whole length of the joint. A 4-mm electrode is recommended for the first or root deposit, with amperage at the upper end of the recommended current capacity (see Fig. 2.21).

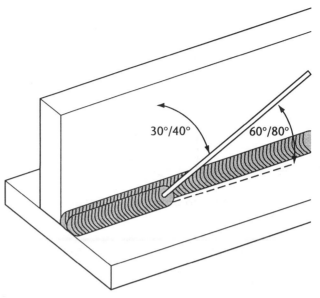

Figure 2.22 Electrode angles for multi-run 'T' fillet joint: second run.

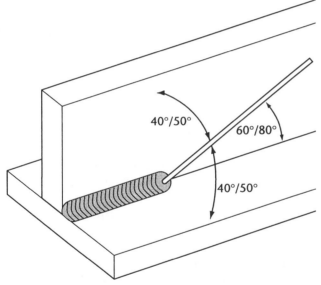

Figure 2.21 Electrode angles for 90° 'T' fillet joint; first run.

Where multi-run fillets are required to achieve the desired weld size, the electrode angles will vary slightly, the second run will have a tilt angle of 30–40° to the vertical plate, the second weld deposit positioned so that it is above the centre line of the first or root run. To achieve this, a slight weave may be required. Slope angles remain the same as for the first or root run in the joint (see Fig. 2.22).

The third run in the sequence is deposited equally between the centre line of the second run and the vertical plate surface. To get the necessary weld shape the tilt angle will be equal between each side of the electrode and the plate surfaces. A slight weave may be

required, but should be kept to a minimum, as there is a risk of slag entrapment in any groove which may have formed between the surface of the first weld deposit and the upper edge (toe) of the second weld deposit. Slope angle will remain the same as for the other deposit in the joint (see Fig. 2.23).

In the horizontal vertical corner joint (10 mm), as shown in Fig. 2.24, the weld sequence will be the same as for the fillet welded 'T'

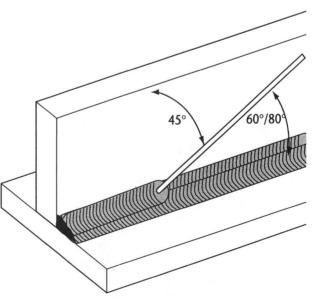

Figure 2.23 Electrode angles for multi-run 'T' fillet joint: third run.

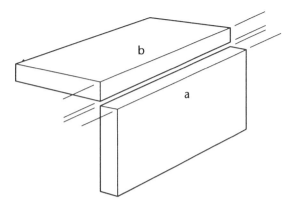

Figure 2.24 Electrode angles for outside corner joint (90°).

joint, although great care has to be taken to avoid rollover on the horizontal edge of the vertical plate (Fig. 2.24a) and burning away of the top corner on the vertical edge of the horizontal plate (Fig. 2.24b). This will result in a misshapen weld profile and a reduction in the throat thickness of the weld, causing weakness of the joint at the corner.

When welding corner joints in the horizontal vertical position it becomes more difficult to maintain the plate thickness on the corner; achieving this continuity of thickness is important to the strength of this joint. The root of first run in this joint also requires a greater degree of control to avoid excessive penetration on the inside of the corner. To get this may require smaller electrodes (2.5 or 3.25 mm) and lower current values (see Fig. 2.25).

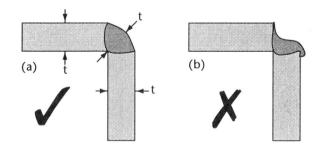

Figure 2.25 Outside corner joint (90°): (a) correct and (b) incorrect.

In order to achieve the correct profile and avoid burning away of the top edge and overlap at the bottom edge, the joint may demand the use of smaller electrodes and lower current values.

Corner joints welded in the flat (gravity) position (10 mm) do not present so much of a problem, although, for example, the size of electrode and current values used will vary in order to control burn-through on the first run.

For the first or root run a 2.5- or 3.25-mm electrode may be required to control the degree of penetration through the joint. Subsequent runs with a 4 or 5 mm electrode may be acceptable, providing no rollover (overlap) is maintained at the top edges (see Fig. 2.26).

Figure 2.26 Open outside corner joint welded in the flat (gravity) position.

A single 'V' butt joint (10 mm) welded in the flat (gravity) position is very similar to the corner joint above, only the joint preparation is different. Instead of being a naturally formed 90° of the square edge plates, the plate edges of the 'V' butt are machined or flame cut to 30–35° to form an included angle of 60–70°. Where the degree of penetration through the joint (welded from one side) has to be controlled, a small electrode (2.5 or 3.25 mm) can be used (see Fig. 2.27).

Weaving should not be necessary, although a slight side-to-side motion may be used to control the heat input and therefore penetration. Root face and root gap will also play a

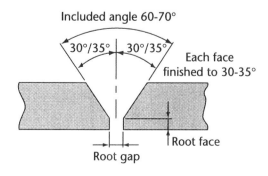

Figure 2.27 Preparation of a single 'V' butt joint.

large part in controlling the degree of penetration through the joint. Increasing the root gap will allow ease of penetration and vice-versa. Increasing the root face will reduce the degree of melting at the root and vice-versa. Subsequent runs can be deposited with larger electrodes, providing that the 'V' is sufficiently wide to allow access of the electrode to any underlying runs. Otherwise fusion penetration will be at risk (see Fig. 2.28).

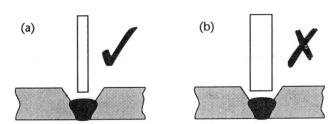

Figure 2.28 (a) Single 'V' butt joint: first run – (a) correct and (b) incorrect. Electrode too large.

There is also the added risk that using too large an electrode with high current values will burn through the first or root run causing excessive penetration or even total collapse of the root deposit. Slope angle for this joint will be 60–80° to the line of the weld and the tilt angle equal to each side of the electrode. The appearance of the finished weld should be blended smoothly with the base metal at the weld toes, with even ripples and no excess metal.

Horizontal–vertical position

Single 'V' butts joint (10 mm) welded in the horizontal–vertical position have a different plate edge preparation, although the included angle will be the same as in the flat 'V' butt (see Fig. 2.29).

Depositing the root or first run in the joint will require the same degree of control as for the flat 'V' butt, although care must be taken to prevent the molten metal sagging away from the top edge. Some degree of support will be given by the lower joint preparation, but in addition to this the electrode should be directed slightly toward the top edge. Other methods of control are the use of smaller diameter electrodes with lower current values. Bead techniques are usually used in this position;

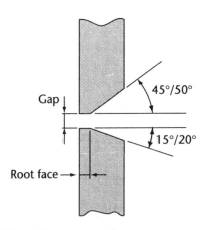

Figure 2.29 Joint preparation.

weave techniques demand high skill levels of the operator.

Weave patterns possibly unique to this welding position are to weave back towards the top edge of the joint, away from the direction of travel. This means that the weld metal on the bottom edge of the joint slightly precedes that at the top, affording some degree of support. Pausing slightly at the top edge will ensure complete fill up of the joint. Stops and starts should always be made on the bottom edge to avoid slag traps and manipulation problems (see Fig. 2.30).

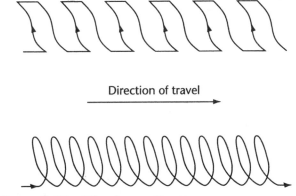

Direction of travel

Figure 2.30 Weaving patterns.

Larger electrodes can be used for subsequent deposits, remembering the point made earlier about access by the electrode to the first and any underlying deposits. When depositing the final layer or capping runs, care must be taken to avoid the molten metal from rolling over onto the bottom face of the plate and burning away the top edge, causing undercutting (see Fig. 2.31).

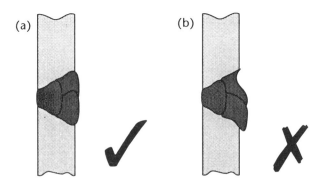

Figure 2.31 (a) Correct profile and (b) incorrect profiles.

The slope angle will beas for other joints listed above 60–80°, while the tilt angle will vary to suit the positioning of the weld deposits in the 'V'. See Fig. 2.32.

Vertical position

Joints welded in the vertical position require the use of the weave technique to a greater extent than some of the earlier examples. The various weave patterns are mentioned above, and there is no hard and fast rule as to which one is used; this will probably be dictated by the joint type or nothing more than the personal choice of the operator. Whatever pattern is chosen will require time to perfect – especially for the beginner.

One of the difficulties to overcome is the control of the molten weld pool, to prevent sagging and the excessive convex weld profile of the finished weld. One of the major influences is the heat input, which is closely linked to the size of the electrode and the current used. Using the weave technique will also influence the amount of heat at any one point at any one time, i.e. moving the electrode within the confines of the joint will help spread the heat. This is common to all vertical welds such as corners, butts and fillets.

Vertical 'T' fillet

Achieving and controlling penetration either into or through the root of the joint is the main concern in all situations. In the case of the vertical 'T' fillet (10 mm) the root of the joint is known as closed (no gaps) and therefore any obstruction, such as slag, between the root of

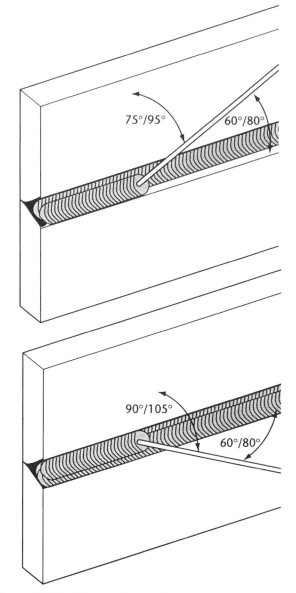

Figure 2.32 Electrode angles.

the joint and the electrode will prevent penetration taking place. It is for this reason that when depositing the root run, the electrode angle must not be allowed to be too steep, certainly no more than 5–10° below 90° to the joint. If the angle becomes too steep then there is a tendency to prevent the molten flux coating (slag) flowing from the joint, or even pushing it up into the corner and trapping it under the weld, forming slag inclusions and/or lack of penetration at the root (see Fig. 2.33).

Commence to deposit the root run starting from the bottom of the joint; weaving should be kept to a minimum, but a slight inverted 'V'

(a)

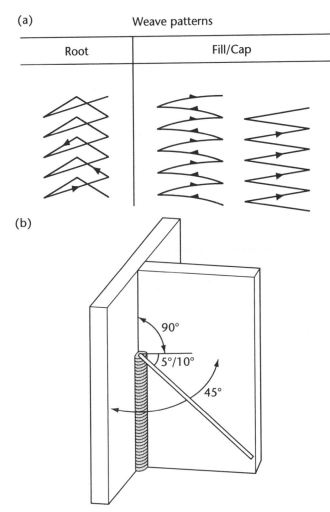

Figure 2.33 Vertical 'T' fillet; (a) weave patterns and (b) electrode angles.

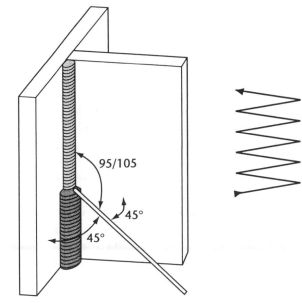

Figure 2.34 Vertical 'T' fillet.

motion will reduce the convexity of the root deposit (flatten it out). For subsequent runs the weave pattern should be chosen to achieve the desired leg length (in this case 10 mm) and weld profile. The slope angle as mentioned earlier should be no more than 5–10° below a line 90° to the joint; the tilt angle equal to each side of the electrode and the plates (see Fig. 2.34).

Open outside corner joints

Open outside corner joints welded in the vertical position (10 mm) are much the same as the vertical 'T' fillet, in that the weld is deposited between two 90° faces – but this time controlling penetration through the joint. This can be achieved by using small

electrodes of 2.5 or 3.25 mm diameter; above this will involve higher current values, causing difficulties with penetration and weld pool control.

It is also important not to burn away the two outer corners of the plate edges and for this reason it may be necessary to restrict the size of the electrode used for the capping run. Although a weaving technique is used to control the sagging of the metal, large diameter electrodes, high heat inputs and therefore large weld pool size will result in sagging of the molten metal, convex weld shape, misshapen weld profiles and excessive burning away of the corners giving rise to undercutting (see Fig. 2.35).

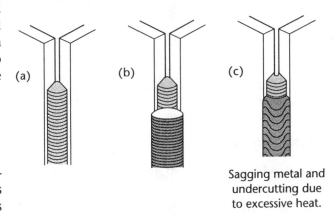

Sagging metal and undercutting due to excessive heat.

Figure 2.35 Vertical outside corner joints; (a) root run; (b) capping run – correct profile; (c) capping run – incorrect profile.

Vertical 'V' butt welds

For vertical 'V' butt welds (10 mm) the technique will be similar to the vertical corner joint. This time the plate edge preparation will form a 60–70° 'V', rather than a 90° 'V', and the problem of the plate corers has been eliminated, although the same care will be required when capping to avoid undercutting at the weld toes.

As mentioned in earlier examples with 'V' butt preparations, some degree of penetration control can be achieved by varying the root gap and face, but this has limitations. Achieving acceptable penetration will require the use of smaller diameter electrodes as in vertical corner joints. Subsequent runs can be deposited by using 3.25 or 4 mm electrodes. Remember the points made earlier regarding electrode access to the joint and underlying deposits, and also the effects of large diameter electrodes, high heat inputs and large weld pool sizes (see Fig. 2.36).

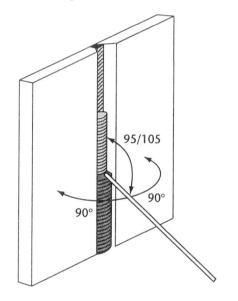

Figure 2.36 Electrode angles for vertical single 'V' butt joint.

Overhead 'V' butt welds

Overhead 'V' butts (10 mm) make excessive demands on the skill of the operator to overcome the gravitational forces and produce acceptable welds. In this position the root face and gap will play an important role along with the size of the electrode and heat input in the

quick freezing of the deposited weld metal to keep it in position. Initial difficulties may be experienced achieving a penetration bead on the upper surface of the joint. Depositing the root run will require a 2.5–3.25 mm electrode slope angle of 60–80° to the line of the weld – with the end of the electrode right up into the 'V' at the front edge of the deposit. This will assist in the fusion of the plate edge and give a slight protrusion through the joint onto the upper face. For subsequent runs 3.25 or 4.00 mm electrodes may be used, maintaining the same slope angle and keeping weaving to a minimum. Although weaving the cap is possible it can be deposited in bead formation rather than a weave over the full width of the joint. This will help control the sagging of the molten metal (see Fig. 2.37).

The arc length must be kept short at all times, caution being taken to avoid the electrode freezing in the weld pool, the root or sides of the joint. Welding in this position is most difficult and may require lots of practice.

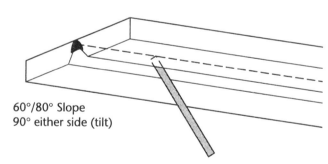

60°/80° Slope
90° either side (tilt)

Figure 2.37 Electrode angle for overhead 'V' welding.

Overhead 'T' filters

Overhead 'T' filters (10 mm) are to some extent a repeat of the horizontal vertical 'T' fillet, flat position for both root and subsequent runs. As with all square edge 'T' fillet joints the mass of parent metal at the root is far more than with the 'V' butt joint. This will permit larger electrodes to be used with higher current values in order to achieve fusion penetration.

For the root run, keeping a short arc with a slope angle of 70–80° and a tilt angle equal to either side of the electrode, commence welding at one end of the joint. Normal procedure for subsequent runs is to position the runs from the bottom upwards (see Fig. 2.38). Therefore the second run will overlap the root run above its centre line and onto the face of the vertical plate with a tilt angle of 30–40° to the vertical plate keeping weaving to a minimum. The third run is positioned between the peak of the second run and the top (horizontal) plate with a tilt angle approximately equal to either side of the electrode.

Successful welds in the overheat position require careful choice in the size of electrode and current values, as well as the skill of the operator.

As weaving has to be kept to a minimum, to avoid sagging, weld sizes are achieved by the multi-run technique, so in some instances it may be necessary to repeat the sequence of runs.

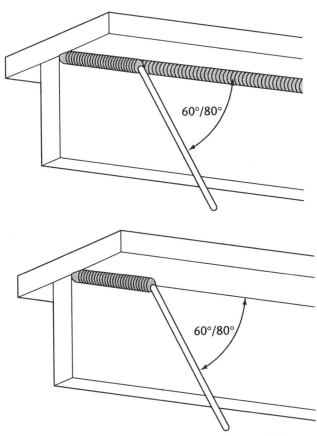

Figure 2.38 Electrode angles for overhead 'T' fillet joints.

Summary

In all of the exercises above a plate thickness of 10 mm was chosen. When welding other thicknesses, first consider all the points that have been discussed in relation to the joint type, plate thickness, welding position and weld size. This will then help the welder decide on:

a) electrode size and heat input in relation to plate thicknesses
b) size of root gap and root face for heat transfer and penetration
c) single bead or weave technique to overcome gravitational forces
d) slope and tilt angles of electrode in relation to deposit
e) positioning of the deposit in the 'V' or corner
f) number of runs required to achieve weld size
g) any difficulties associated with the joint as with corner joints
h) access of the electrode into the 'V'
i) finished weld acceptance levels.

As mentioned at the beginning of this chapter, the stick welding process is a very versatile, well established, mobile and proven process, which is widely accepted by both operator and manufacturer. It is a very mobile process used for welding in all positions, and although more suitable processes are available, it is able to be used for joining most metals in the hands of a skilled operator. Although the process has been around for a long time, it still finds a place in a wide range of workshops and site situations, and although used to a lesser degree than in the past, should still be around for many years to come.

3

Metal arc gas-shielded welding

Choosing the name to describe this process was not without consideration. This type of welding is often referred to by many different names. The reason for choosing **metal arc gas-shielded welding** (MAGS) was that it encompasses the process without regard to other influencing factors such as inert or active gas, solid, cored or self-shielding wires. Other names the process is referred to are: metal inert gas (MIG) welding, metal active gas (MAG) welding, gas metal arc welding (GMAW), CO_2 welding and semi-automatic welding. To satisfy the training situation the description MAGS is used in the text below. You will find that the process is most commonly referred to as MIG or MAG in the workplace.

In order to comply with international standards a listing of the processes and their numerical representations is given in Appendix 1 at the end of the book.

MAGS welding process

The MAGS welding process (shown in Fig. 3.1) can be used on a wide range of metals, positions and thicknesses. The process works by creating an arc between a continuous wire electrode and the parent metal to be joined. The wire feed speed unit controls the rate at which the wire is fed to the arc. Increasing the wire feed speed will proportionately increase the current (amperage) on the wire. For example, a 1.6 mm diameter wire will operate over a current range of 150–400 amps with a proportionate wire feed

speed increase from 1.9 to 7.6 m/min. Therefore, over the wide range of wire sizes available, e.g. 0.8, 1.00, 1.2 and 1.6 mm, the process is suited to a wide range of current values, metal thicknesses and welding positions.

A power source with a DC output and the electrode (wire) connected to the positive pole supplies current to the wire via the contact tip in the welding torch. Current, gas, wire and water (in the case of water-cooled torches) are supplied to a hand-held torch by means of a flexible conduit. Voltage and wire feed speed (amps) are adjusted at the power source to suit welding conditions, e.g. wire diameter, welding position, metal thickness, mode of transfer and type of wire. The welder is then left to control the distance between the torch and the work surface and the speed of travel.

As mentioned earlier, the end of the molten wire electrode, the transferring metal, weld pool and the molten parent metal must be protected from atmospheric contamination. In the MAGS welding process this is achieved by the gas shield, not from the electrode but from a separate gas supplied in cylinders or bulk supply piped to the power source. The level of shielding provided by the various gases will vary with the density of the gas; in general terms, the heavier the gas the better the shielding, especially in the flat position. Argon, carbon dioxide and oxygen are the heavier of the common shielding gases.

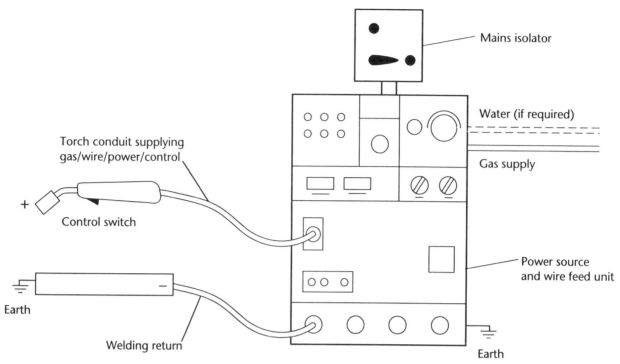

Figure 3.1 Metal active gas welding (MAGS) equipment.

Choice of shielding gas

The shielding gas must be carefully chosen to suit the application. The selection will depend on:

a) The compatibility of the gas with the metal being welded.
b) Physical properties of the material.
c) The welding process and mode of operation.
d) Joint type and thickness.

If the material has a high thermal conductivity a shielding gas which increases the heat transferred to the work piece should be used. For copper and aluminium, helium or helium/argon mixtures are particularly useful since they reduce preheat requirements and improve penetration on thicker sections. A summary of common gases and their application is given in Table 3.1.

Carbon dioxide (CO_2) supplied in liquid form tends to cause freezing up of the regulator during use as a result of the change to a gaseous form. When using CO_2 for gas-shielding purposes a heater/vaporiser is fitted between the cylinder outlet and the regulator. The heater should be allowed to warm up before welding commences. A flowmeter can also be incorporated on the outlet side of the regulator, if required (see Fig. 3.3).

Electrode wires

Welding wires are usually supplied on 12–18 kg reels layer wound in such a way as to ensure that the wire releases smoothly during the welding operation. Steel wires are usually coated with copper to offer some protection against corrosion, to reduce friction and to improve the electrical contact between the wire and the contact tip. Not all wires are copper coated; these are usually used where the rate of wire consumption is high and corrosion resistance is not a problem. Both types of wires should be stored in their original packaging in a dry environment until needed.

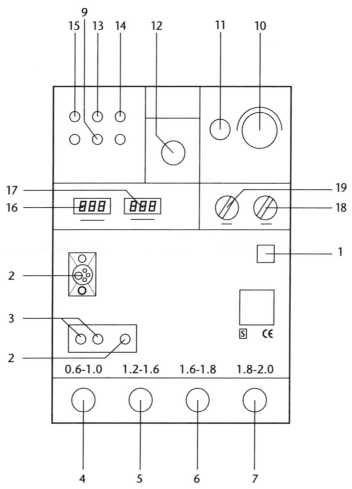

Figure 3.2 MAGS welding equipment (courtesy of Migatronic). (1) Main switch. (2) Central adapter – for welding torch. (3) Quick release connector – for the torch cooling system. (4) Inductance output – Ø0.6–1.0. (5) Inductance output – Ø1.2–1.6. (6) Inductance output – Ø1.6–1.8. (7) Inductance output – Ø1.8–2.0. (8) Connection for push–pull hose and torch assembly. (9) On – lights when the machine has been turned on. (10) Wire speed – this button is used to set the wanted wire speed. Adjustable from 1.7 to 24 m/min. (11) Trigger mode – this switch is used for setting of either 2-stroke or 4-stroke trigger function. (12) Inching button – this button is used for fitting of welding wire. When the button is activated the wire is fed by the speed selected by the wire speed button (position 10). Inching can also be done by means of the trigger of the torch handle by activating it for at least 3 s. Afterwards the wire will inch by the speed selected on the wire speed button (position 10). (13) Burn back – adjustable delay from the wire feed stops until the welding voltage is out in order to avoid sticking of the wire. Adjustable between 0.05 and 1 s. (14) Gas post-flow – gas post-flow time. Adjustable between 0 and 20 s. (15) Soft start – pre-setting of soft start which means the speed by which the wire starts before the arc is ignited. Adjustable between 1.7 and 5 m/min. In position OFF the wire will start by the speed selected by the wire speed button (position 10). (16) Voltmeter – shows the welding voltage. (17) Ammeter – shows the welding current. (18) Fine adjustment of welding voltage. (19) Coarse adjustment of welding voltage.

Wire feed system

The wire feed (push) system is shown in Fig. 3.4. The electrode wire is fed from the reel to the tube via the wire feed rollers. Pressure on the rollers can be adjusted by means of a tension screw. Excessive pressure on the rollers may distort or damage the wire, causing feeding problems in the lines or at the contact tip. This system is known as the push system. In

Table 3.1 A summary of common gases and their applications

Gas	Applications	Features
CO_2	MAGS plain carbon and low-alloy steels	Low-cost gas. Good fusion characteristics/ shielding efficiency but stability and spatter levels poor. Normally used for dip transfer only
Argon + 1–7% CO_2 + up to 3% CO_2	MAGS plain carbon and low-alloy steels. Spray transfer	Low heat input, stable arc. Finger penetration. Spray transfer and dip on thin sections. Low CO_2 levels may be used on stainless steels but carbon pick up may be a problem
Argon + 8–15% CO_2 + up to 3% CO_2	MAGS plain carbon and low alloy steels. General purpose	Good arc stability for dip + spray pulse and FCAW. Satisfactory fusion and bead profile
Argon + 16–25% CO_2	MAGS plain carbon and low alloy steels. Dip transfer /FCAW	Improved fusion characteristics for dip
Argon + 1–8% O_2	MAGS dip, spray and pulse, plain carbon and stainless steel	Low O_2 mixtures suitable for spray and pulse but surface oxidation and poor weld profile often occur with stainless steel. No carbon pick up
Helium + 10–20% argon + oxygen + CO_2	MAGS dip transfer, stainless steel	Good fusion characteristics, high short circuit frequency. Not suitable for spray/pulse transfer
Argon + 30–40% helium + CO_2 + O_2	MAGS dip, spray and pulse welding of stainless steels	Improved performance in spray and pulse transfer. Good bead profile. Restrict CO_2 level for minimum pick up
Argon + 30–40% helium + up to 1% O_2	MAGS dip, spray and pulse welding of stainless steels	General-purpose mixture with low surface oxidation and carbon pick up. (It has been reported that these low oxygen mixtures may promote improved fusion and excel lent weld integrity for thick section alu minium alloys)

Gas mixtures available and their applications (courtesy of BOC)

some cases the feed rollers are mounted on the welding torch feeding the wire only a short distance. This system is known as the pull system. A combination of both these systems can be used where feeding difficulties may occur.

Metal transfer modes

The MAGS welding process offers three methods of transferring metal across the arc, these are:

■ dip or short circuit transfer
■ spray transfer
■ pulsed transfer.

Dip transfer takes place at 15–25 volts and 40–200 amps on wire diameters up to 1.2 mm. This mode of metal transfer is suitable for welding thin sheet, positional welding of thicker sections and depositing root runs in open butt joints. Metal transfer is a result of reducing the arc gap until the wire comes into contact with the work or molten pool, i.e. dip-

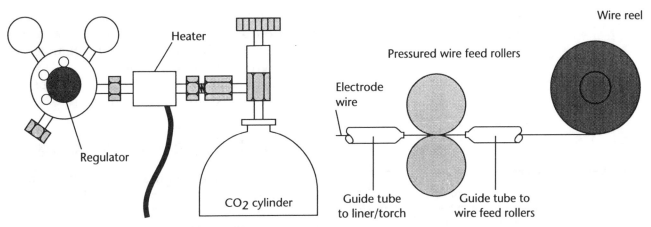

Figure 3.3 Special arrangement for CO_2.

Figure 3.4 MIG electrode wire feed system.

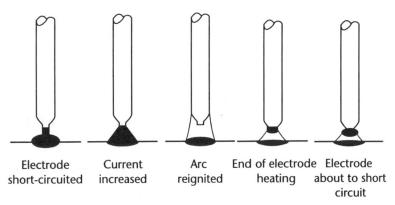

Figure 3.5 Dip transfer.

ping or short circuiting. At the point of short circuiting a rise in current takes place, melting off the end of the wire and reigniting the arc, the cycle is repeated 100–200 times per second. If the rise in current during short circuiting is too fast it will cause the molten globule to explode out of the weld pool causing excessive spatter. If the current rise is to slow the wire will stub and possibly freeze into the weld pool. This can be controlled by the inductance setting, i.e. the higher the inductance setting, the lower the short circuiting occurrence and vice-versa (see Fig. 3.5).

Spray transfer takes place at 30–50 volts and 200–400 amps on wire diameters over 1.00 mm. This mode of metal transfer is suitable for welding metal thicknesses above 6 mm with a high rate of deposition and good penetration and a high heat input. Welding is limited to butts and horizontal vertical fillet welds in the flat position. The exception to this is the welding of aluminium and its alloys which can be

welded in position. This is possible due to the oxide film which forms on the surface of the weld pool retaining the molten metal in position. Metal transfer is a result of the increased current values. At higher current values the globule size decreases and the rate of metal transfer increases. This results in the molten metal being projected across the arc in small droplets in the form of a fine spray with low spatter levels. The high welding speed, rate of metal deposition, size of deposits in one pass and good penetration capabilities makes it an economical method of welding heavy steel sections (see Fig. 3.6).

Pulsed transfer has gone some way to combine the two modes above by controlling the melting off period in which the droplets are detached from the electrode. This is made possible by introducing a high current pulse into the circuit, at which stage the droplet transfer takes place whilst a low background current maintains the arc. The result is a spray mode of

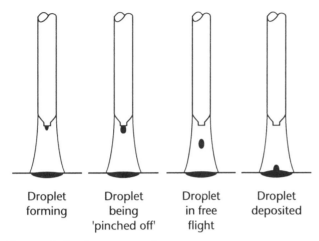

Figure 3.6 Spray transfer.

metal transfer at lower average current values, enabling positional welding at higher deposition rates, good fusion in open butts with controlled penetration capability, reduced spatter levels and good arc stability (see Fig. 3.7).

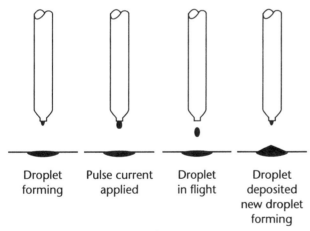

Figure 3.7 Pulsed transfer.

The self-adjusting arc

In the MAGS welding process the voltage setting of the power source governs the arc length. Small diameter wires up to 1.6 mm burn off at very high rates. The burn-off rate will vary with small changes in current caused by variation in arc length due to torch movement, erratic wire feed or fluctuating current pick up in the contact tip. A power source which keeps a moderately constant voltage over a wide current range (constant potential or flat output power source) provides a means of controlling the arc. Any small changes in arc length are rapidly made up for by an increase in current (wire feed) which keeps the arc length constant, i.e. self-adjusted.

Contact tip, nozzle settings

The position of the contact tip in relation to the nozzle is slightly different for the modes of metal transfer. The contact tip can be set to protrude beyond the nozzle or set back inside depending on whether using dip or spray transfer mode. This will affect visibility, accessibility and gas shielding. The following gives suggested settings for the mode of metal transfer being used.

Mode of transfer	Contact tip
Dip	2–8 mm protruding
Spray	4–8 mm inside
Spray (aluminium)	8–10 mm inside

Electrode extension

The extension of the wire or stick out is usually measured from the contact tip to the weld pool. Excessive stick out will reduce the arc current resulting in less penetration. Suggested electrode stick out for the respective modes of transfer are shown below.

Mode of transfer	Electrode extension
Dip	4–8 mm
Spray	15–25 mm

Burn back

The result of a burnback is fusing of the electrode wire to the end of the contact tip. The remedy can be quite simple in some cases, not so in others depending on the severity of the burnback. It may be possible to release the wire by pulling it free with a pair of side cutters; lightly filing the end of the tip may also be helpful. If this is successful the wire will spring forward on release. If this is unsuccessful switch off the power source and remove the contact tip to expose and cut the wire. The wire can then be gripped in the vice and by gently turning the contact tip the wire should release. In severe cases it may be necessary to cut off the end of the contact tip to release the wire or even renew the contact tip.

Current and voltage settings

The following is a guide to voltage settings for MAGS welding. Slight adjustments will be required to suit plate thicknesses, joint type, joint design, welding position, root face, type of material and procedure demands.

Wire diameter (mm)	Voltage (volts)	Current (amps)
0.6	15–22	50–100
0.8	18–25	60–190
0.9	18–30	70–250
1.0	18–32	80–300
1.2	18–36	120–400
1.6	27–40	220–500

British Standards for wires and fillers

BS2901 – Part 1 Filler Rods and Wires for Gas Shielded Arc Welding of Ferritic Steels covers the chemical composition, diameter and tolerances of rods and wires, condition of rods and wires, dimensions of reels of wire, packaging and marking.

Features of the process

Some of the more important features of the process are summarised in Table 3.2.

Weaving

Weave patterns are common to all the processes under discussion. In MAGS welding, weaving usually refers to the side-to-side movement of the torch and, as a result, the electrode. In this section the weave patterns, the reasons for weaving and the associated problems are considered.

The MAGS welding process is inclined to suffer from 'lack of side-wall fusion', pene-

Table 3.2 Some of the more important features of the process (courtesy of BOC)

Feature	Comment
Low heat input	All modes of transfer, particularly dip and pulse give low heat inputs compared to the MMA process. This is useful for thin plate and positional work but care must be taken to avoid fusion defects on thicker material
Continuous operation	High operating efficiency. High duty cycle. High productivity
High deposition rate	Higher deposition rate compared to the MMA process particularly with spray transfer
No heavy slag	The absence of slag means that little cleaning of the weld is required after welding
Low hydrogen	Absence of flux coating and the use of dry gas controls the hydrogen level. This reduces the risk of cold cracking

tration into fusion faces on heavy sections or wide preparations. Weaving the electrode is a means of overcoming this, ensuring the full force of the arc is directed toward the fusion faces and any underlying weld deposits to ensure good fusion penetration. This may influence the choice of weave pattern chosen.

Stops and starts

The same rule applies as for MMA welding. Strike the arc in advance of the previous weld and within the confines of the weld joint, moving as quickly as possible into the weld crater. Stops and starts should blend smoothly without any abrupt change in weld shape.

On completion of the weld it is important to fill the weld pool (crater) to ensure continued weld profile, throat thickness, strength, etc. When crater fullness has been achieved and the arc extinguished, the torch should be held at that position, at the end of the weld to provide gas shielding to the cooling weld metal, avoiding the risk of contamination by atmosphere.

Slope and tilt angle

Although the angles are the same as MMA welding the stance of the operator may change with the MAGS process. This is due to the large diameter of the gas nozzle restricting visibility to some degree. To overcome this the operator will have to adopt a stance which permits good visibility of the nozzle-to-work distance, joint line and weld pool behaviour. Reducing the slope angle will enable better visibility of the weld area although this may have a marked effect on the level of spatter, penetration, and gas shielding.

Direction of travel

The technique used may be referred to as leftward or rightward, backhand or forehand but possibly the most common to the MAGS welding process is pushing or pulling. Both techniques have their relevance in the process and should be practised by the new operator. Not forgetting what was said earlier about restricted visibility, pushing is possibly the most widely used, especially on thicker sections. The pulling technique, although not the most popular, is particularly suited to vertical down welding of thin sheet and root runs in open butt joints (see Fig. 3.8).

Flux-cored arc welding

Flux-cored arc welding (FCAW) is a variation of the MAGS process. It forms an arc between a continuous wire electrode and the part to be welded, forming a weld pool and melting of the joint faces takes place to produce a welded joint. The process uses a gas shield produced by the flux contained within a tubular electrode. It can be used with or without a supplementary shielding gas.

The flux-cored electrode is a made up tubular filler metal electrode consisting of a metal

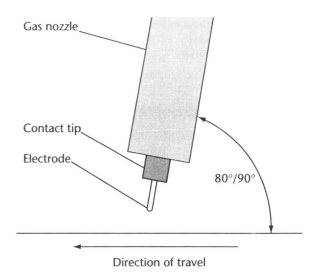

Figure 3.8 Electrode and nozzle arrangement for MIG welding.

outer sheath and a core of various powdered materials (see Fig. 3.9). During the welding process an extensive slag covering is produced on the face of the weld bead. The feature that distinguishes the FCAW process from other arc welding processes is the enclosure of the fluxing ingredients within the electrode.

The continuously wire-fed electrode has remarkable arc operating characteristics. FCAW offers two fundamental process variations that differ in their method of shielding the arc and

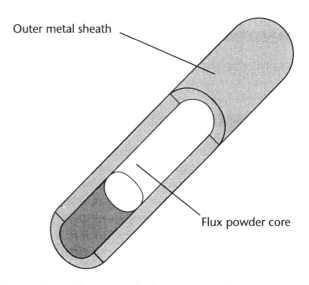

Figure 3.9 Flux-cored wire construction.

the weld pool from oxygen and nitrogen atmospheric contamination. One type, self-shielded flux-cored arc welding, protects the molten metal by vaporisation of the flux core in the heat of the arc. The other type, gas-shielded flux-cored arc welding, uses an additional protective gas flow to the flux core. With both methods the core material provides a substantial slag covering to protect the deposited weld metal during solidification.

The flux cored arc welding process combines three major features of other arc welding processes. These are:

a) the productivity rates of continuous wire electrode
b) the metallurgical benefits of the flux composition, and
c) a slag that supports, protects and shapes the weld deposit.

This combines the operating characteristics of the manual metal arc (MMA), metal inert/active gas (MIG/MAG) and the submerged arc welding (SAW) processes. As well as the features listed above the two processes shown below are the self-shielded flux-cored arc welding process and the gas-shielded flux-cored arc welding process.

Self-shielded flux-cored arc welding

Figure 3.10 shows that the shielding is obtained from the vaporised flux ingredients which displace the air and protects the molten metal droplet during transfer. The vaporised flux then solidifies to protect the weld during solidification and assists in retarding the cooling rate of the deposited metal.

One characteristic of the self-shielded process is the use of long electrode extensions (electrodes extending beyond the contact tube). Electrodes of 19–95 mm are not unusual depending on the application. This preheats the electrode and lowers the voltage drop across the arc at the same time the welding current decreases, lowering the heat input. The result is the weld bead being narrow with shallow penetration making the process suitable for welding light-gauge materials (see Fig. 3.11).

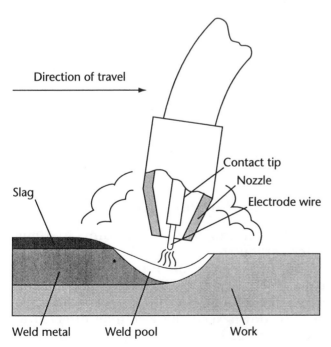

Figure 3.10   Self-shielded flux-cored arc welding.

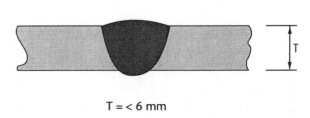

T = < 6 mm

Figure 3.11   Single butt joint in thin plate; suited to gas-shielded flux-cored arc welding.

Gas-shielded flux-cored arc welding

Gas-shielded flux-cored arc welding (GSF-CAW) is illustrated in Fig. 3.12.

The gas-assisted method, the shielding gas, either carbon dioxide (CO_2) or a mixture of argon plus CO_2 protects the molten droplet from atmospheric contamination by forming an envelope around the arc and over the weld pool. Some oxygen may be generated by the break up of the shielding gases, but the composition of the electrode flux is formulated to provide deoxidisers to eliminate any small amount of oxygen in the gas shield.

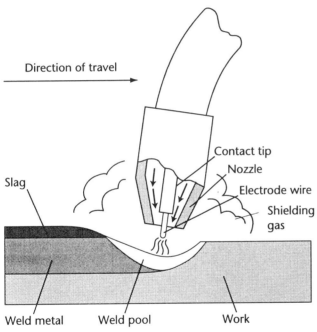

Figure 3.12   Gas-shielded flux-cored arc welding.

MAGS positional welding

Positional welding with MAGS follows many of the same rules as discussed in the section on position welding with MMA. Where possible, the work should be manipulated or rotated so that welds can be made in the flat position, the molten metal being pulled into the joint by gravitational forces. As mentioned earlier this is not always possible.

Both vertical up and vertical down welding techniques have a wider application with this process than with MMA. The slope angle will vary from 60° to 80° in the direction of travel for vertical down welds on thin sheets and root runs in open butt joints to 90–120° in the direction of travel for vertical up filling runs in butt joints and root and subsequent runs in fillet welds. The tilt angle in most cases will be equal to either side of the torch and the plate surface.

Although it was mentioned earlier in this section that the pulsed transfer mode of MAGS welding allows positional welding at higher average current values and the added advantages, the initial cost of this equipment is very expensive. This, along with other reasons, is why the dip or short circuiting mode of metal transfer is commonly used for positional welding. With the use of wire diameters up to 1.2 mm for vertical (up and down) and overhead positions, it is possible to make quick freezing deposits as a result of the small weld pool size.

Although the vertical down technique is common practice for depositing runs at various stages on butts and fillets of different thicknesses, care must be taken to avoid molten metal running in front of the weld pool preventing access to the root of the joint.

When discussing the welding positions below it would be impractical to state precise welding currents and voltages for any particular position. For this reason the current and voltage settings given below will be influenced by plate thickness, welding position and wire diameter. They offer a starting point to which final tuning will be required to achieve the ideal welding conditions. Other factors affecting these conditions are heat sinks from backing bars, jigs or adjoining plates, joint types, joint preparation and type of material being welded. All will have to be considered.

'T' fillet and square edge butt joints

'T' fillet and square edge butt joints in 3-mm thick carbon steel, vertical down. Slope angle 60–70° in the direction of travel and tilt angle equal to either side of the torch, using 0.8 mm wire at 19–21 volts and 90–130 amps. Welding vertical down keep the electrode directed at the leading edge of the weld pool, taking care to avoid wire protrusion through the gap (butt joint). Speed of travel will be dictated by the rate of metal deposition. Tack welds should be of sufficient size, strength and quantity to prevent movement during welding (see Fig. 3.13).

Horizontal–vertical 'T' fillets

Horizontal–vertical 'T' fillets in 3-mm thick carbon steel, flat position. Slope angle 100–120° in the direction of travel and tilt angle equal to either side of the torch, using 1.0–1.2 mm wire at 18–25 volts and 120–200

amps weld from right to left (left to right for a left-handed person) using the push technique deposit the root weld with minimum weave directing the full force of the arc at the root of the joint. Where it is necessary to use multi-runs, for example, on heavier sections slope and tilt angles will change with the respective run. Examples are shown in Fig. 3.14. Tack welds should be of sufficient size, strength and length to prevent movement during welding. Note, if positioning equipment and joint type allow, larger diameter wires or flux-cored metal wires can be used to achieve heavier weld deposits.

Single 'V' butt

Single 'V' butt 10-mm thick carbon steel, vertical position, vertical down root, vertical up fill and cap, 1.6–3.0 mm root gap and 1.0–2.0

Root gap 1.0 to 1.6 mm minimum weave to ensure fusion of both edges

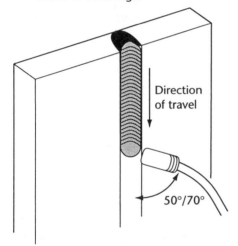

Keep weaving to a minimum, a slight side to side inverted 'U' weave is acceptable

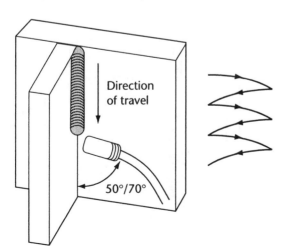

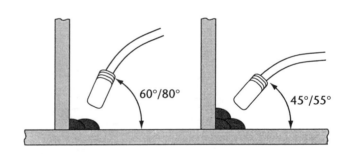

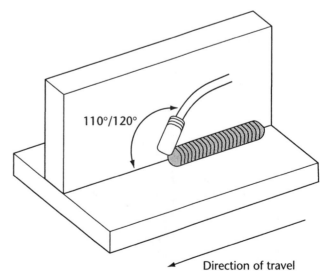

Figure 3.13   Vertical fillet and butt welds.

Figure 3.14   'T' fillet weld in flat position.

mm root face with a 60° included angle. Using 1.0–1.2 mm wire at 18–24 volts and 90–240 amps. Minimum weaving will be required on the root run, the wire directed at the leading edge of the weld pool ensuring penetration and fusion of both edges. Weaving will be required for filling and capping; suggested techniques are shown in Fig. 3.15, along with the slope and tilt angles for the direction of travel.

Thicknesses of up to 20 mm can be accommodated by increasing the number of filler runs. For plate thicknesses over 20 mm different joint preparations (double 'V') or joint design (thicker root face) may allow vertical up root runs.

Fillet weld

Fillet weld in 12–20 mm thick carbon steel plate, vertical position. Using 1.0–1.2 mm wire at 18–24 volts and 105–220 amps weld vertical up, slope angle 105–120° in the direction of travel. Minimum weaving is required for the first run concentrating the force of the arc in the root of the joint. Filling and capping is achieved by weaving subsequent runs (Fig. 3.16a or b) or block weave (Fig. 3.16c), care being taken to avoid excessive heat build up or convexed bead shape.

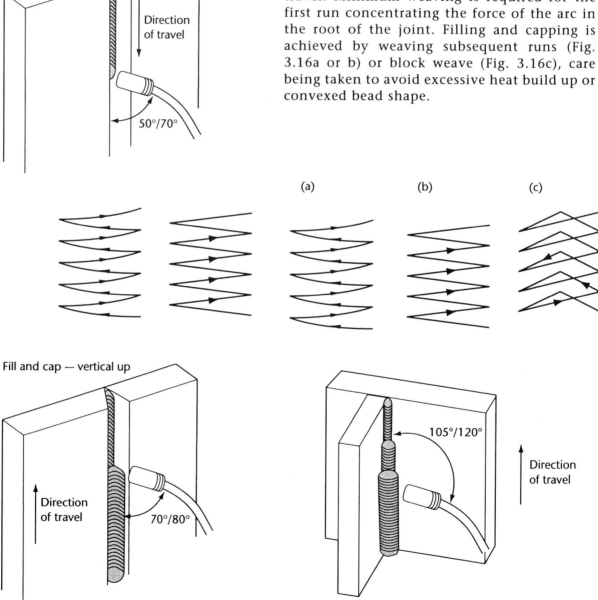

Figure 3.15 Butt welds in thick sections.

Figure 3.16 Fillet welds in thick sections.

Horizontal–vertical butt joints

Horizontal–vertical butt joints in 12 mm and above carbon steel. Using 1.2–1.6 mm wire at 18–22 volts and 140–220 amps. Slope angle 10–120° in the direction of travel. A bead rather than a weave is used in this position to avoid excessive heat build up and sagging (overlapping) of the molten metal resulting in lack of fusion penetration into the fusion faces or underlying weld runs. The number of runs required will be determined by the plate thickness and type of joint preparation (see Fig. 3.17). Weave techniques are possible but demand a high level of skill from the operator. In most instances the arrangement of runs will be as shown, the previous run affording some degree of support for the one above. The tilt angle (to either side of the electrode) will vary with the run being deposited; a good indication of this is the correct position of the weld in the joint preparation and avoiding sagging of the molten metal shown in Fig. 3.18. Recommended tilt angles are shown in Fig. 3.19 for the respective weld beads.

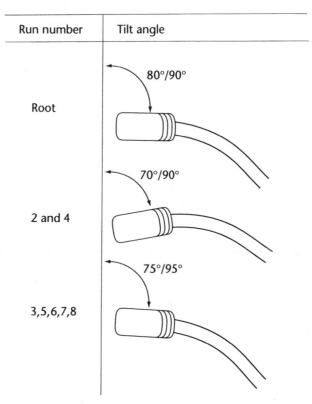

Run number	Tilt angle
Root	80°/90°
2 and 4	70°/90°
3,5,6,7,8	75°/95°

Figure 3.19   Tilt angles.

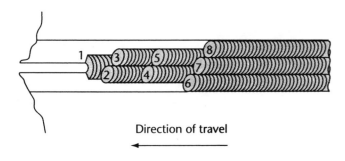

Direction of travel

Figure 3.17   Arrangement of welds runs.

If weaving is required, the pattern should be backwards toward the top edge of the joint. Any weave pattern adopted should be as small as possible in keeping with good weld profile, good fusion, positioning of weld bead and heat build up (see Fig. 3.20).

The welding process is widely used in a variety of industrial situations. Wire diameters of 0.8 mm can be used in sheet metalwork and motor vehicle manufacture and repair where the heat-affected zone is smaller than that of

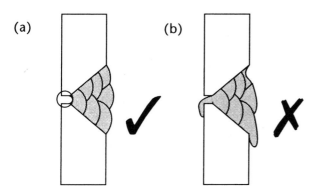

Figure 3.18   Multi-run butt weld: (a) correct profile and (b) incorrect profile.

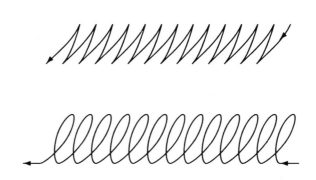

Figure 3.20   Suggested weave patterns.

oxyacetylene welding, thus reducing distortion and increasing welding speeds. Wires of 2.4 mm flux and metal-cored wires can be used for heavier fabrication, e.g. offshore installation manufacture. The process offers competition for the MMA process. When choosing a process, likely considerations are initial cost, mobility of equipment, working environment, cost of consumables, production rates and automation. This list is by no means exhaustive and the decision must be made carefully. MAGS can be used both as a semi-automatic process, manually operated and as a fully automated system with the use of robots. The process uses a continuous wire electrode which eliminates electrode changing and hence the number of stops and starts during welding. By only using a gas shield as a form of arc protection it eliminates the use of fluxes which reduces deslagging and post-cleaning operations of welds where flux residues can cause corrosion problems.

With the wide range of shielding gases available and the use of flux-cored wires, the process has a wide range of applications. The process is suitable for welding mild and low-alloy steels, stainless steels, aluminium and its alloys, copper and bronzes. Used with flux-cored wires it is possible to deposit wear-resisting overlays for the prevention of corrosion, impact and abrasive wear. The process can be used for welding in all positions with the associated advantages. Training of operators is no more difficult or less important than any of the other processes under discussion. The main disadvantages are the initial cost and mobility of the equipment.

Safety precautions to be observed are the same as those associated with the other processes and discussed in Chapter 1. The subject of ventilation and fume extraction must be carefully considered, with the type of shielding gas being used. Bearing in mind that CO_2 and argon gases are heavier than air there is the added risk of the exclusion of oxygen which will result in suffocation or carbon monoxide poisoning. Always ensure that good ventilation and any extraction is correctly sited to remove any harmful gases that may accumulate.

4

Tungsten arc gas-shielded welding

The tungsten arc gas-shielded (TAGS) process is also known as tungsten inert gas (TIG) welding, argon arc welding and gas tungsten arc welding. The reason chosen for using the name TAGS is that it is a good description of the process. The process uses a non-consumable tungsten electrode which forms an arc as a source of heat to melt the work, all being protected by a gas shield.

TAGS welding equipment

The TAGS welding equipment is shown in Figs 4.1 and 4.2. The TAGS process lends itself to welding of both ferrous and non-ferrous metals in a wide range of positions and thicknesses, although on thicker sections (above 8 mm) other processes can be more suitable or economical. The main difference between this process and the other arc processes is that in TAGS welding the electrode is not consumed in making the weld. If required, additional metal can be added to the joint in the form of a filler wire as is the case with oxyacetylene welding.

By making use of both AC and DC electrode diameters of 1.0–6.0 mm and current values ranging from 15 to 350 amps means that the process is suitable for both repair work and manufacture of new installations. Although TAGS welding is relatively slow when compared to other arc welding processes, its ability to produce high-quality controllable welds without the need for fluxes for welding aluminium and stainless steel adds to its accept-ability in a wide variety of industrial applications.

Equipment selection

The power source is usually a transformer/rectifier type with both AC and DC output. With open circuit voltage (OCV) of 80–100 volts, AC will be required for welding of aluminium and its alloys. DC is required for welding carbon alloy and stainless steel, copper and nickel alloys. The current capacity of the equipment is chosen to be compatible with the type of work undertaken. Typical examples are given in Table 4.1 showing current ranges related to plate thicknesses for steel and aluminium.

Welding in the high amperage range for prolonged periods will mean that water-cooling is needed on the equipment. Most modern power sources have built-in, self-contained water-cooling systems. Older power sources may be connected to the water-circulating system in the workshop; water from the mains supply is passed through the torch to a drain. Alternatively, water is supplied from a tank circulated by means of a pump to the welding torch, returning to the tank via a water cooler. The flow of water is governed by the water flow valve and operated by the torch or footpedal switch, which also activates the welding sequence. Water cooling becomes necessary at current values over 150 amps.

The flow of the argon shielding gas is controlled in much the same way, permitting the gas to be shut off when welding is not in oper-

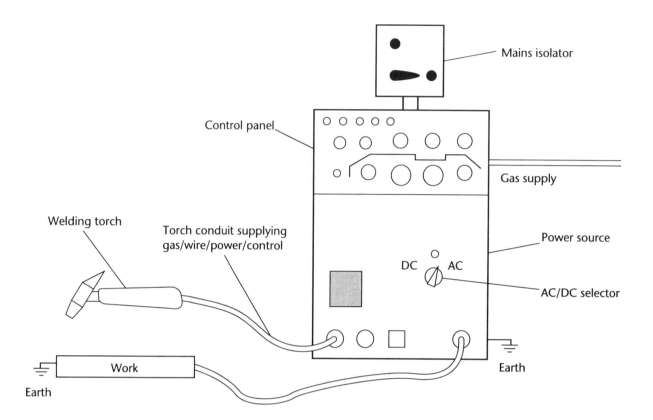

Figure 4.1 Tungsten arc gas shielded (TAGS) welding system.

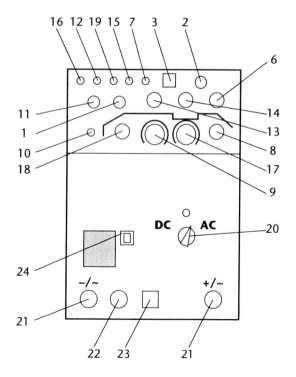

Figure 4.2 AC/DC rectifier: description of the equipment. (1) Function switch. (2) Overload indicator. (3) Ammeter. (4) Green control lamp (not shown). (5) Automatic high frequency (not shown). (6) Arc balance. (7) Post-weld gas flow. (8) Slope down. (9) Current setting. (10) Multiplug. (11) Remote. (12) Hot start. (13) Pulse current time (pulsed arc welding only). (14) Base current time. (15) Pre-weld gas flow. (16) Stop/start current. (17) Reduced current. (18) Slope up. (19) Spot welding time. (20) AC/DC selector. (21) Cable socket. (22) Shielding gas connection. (23) Multi-connect TIG torch control. (24) Voltage safety control.

Table 4.1 Current ranges in TAGS welding

Metal thickness (mm)	Current (amps)	Electrode (mm)
Steels: direct current electrode negative (DC−)		
Up to 2	15–90	1.6
2–5	90–200	2.4
over 5	170–240	3.2
Aluminium: alternating current (AC)		
Up to 1.6	25–85	1.6–2.5
1.6–5.0	80–180	3.2–5.0
over 5.0	160–250	5.0–6.0

ation. Operating the footpedal switch (or torch switch) will allow the gas to flow to the torch giving protection for the electrode, molten pool and filler rod. A gas delay switch is usually incorporated. This allows a predetermined time to be set for the gas to flow after the welding current has been switched off. This is to ensure that the weld metal and the parent metal are protected during cooling. The welding torch should therefore be held at the end of the weld until the gas flow has ceased. This will be the pre-set time, for example, 0–10 s.

Additional components

Additional components making up the TAGS equipment are:

- high-frequency unit or surge injector
- DC suppressor unit
- contactor
- welding torch with conduit (cables and hoses)
- electrodes
- shielding gas.

Surge injector

This provides an alternative means to high frequency (HF) of maintaining arc stability through the zero pauses, without the disadvantage of HF, i.e. electronic equipment interference. The function of the surge injector is to provide a high voltage surge or pulse at the critical period when the negative half cycle of the AC arc changes to the positive half cycle. The positive half cycle assists in surface oxide removal when welding aluminium and magnesium alloys.

DC suppressor

When welding aluminium an arc is formed between two dissimilar metals: aluminium and tungsten. Although when using AC the tendency is for the current to be converted to DC, it is also more noticeable where metals with heavy surface oxide films are being welded. Connecting a DC suppressor into the AC current will allow the passage of AC and effectively block any DC. The suppressor is only necessary when welding with AC and should always be used on high-quality work.

Contactor

The contactor allows the arc to be extinguished without removing the electrode and gas shield from the finished weld. This allows for continued gas shielding of the weld metal until it has cooled. It also provides protection for the operator by switching off the OCV when the torch is not being used. The contractor can be operated by a trigger switch or button on the torch, or by means of a footpedal. The footpedal control also allows current adjustment during the welding operation.

Arc striking

When using the TAGS process problems arise when arc initiation (striking) is achieved by touching the electrode onto the work as this will cause contamination of the work and the tungsten electrode. Transfer of tungsten to the weld (tungsten inclusions) will tend to cause hard spots in the weld, similarly, transfer of parent metals to the tungsten can affect arc stability and contaminate the electrode.

To allow the arc to be struck without the electrodes contacting the work a high-frequency spark ionises the arc gap allowing the pre-set welding current to flow, and the welding to start. When using DC the high frequency is automatically switched off once the arc has been established. If AC is being used the high-frequency facility is left in the circuit and is used to assist re-ignition of the arc at zero current periods during welding. This helps to stabilise the AC arc. This boost takes place at 100 times per second (see Fig. 4.3).

Having this high-frequency facility available means not having the electrode contact the work. This assists in maintaining the shape of the electrode and increases the life expectancy of the electrode. The disadvantage of high frequency is the interference with radio, television, computers and electronic systems nearby. Most modern equipment will be manufactured using one of the alternative methods listed below.

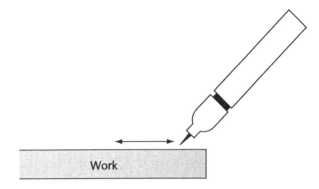

Figure 4.4   Scratch-starting an arc.

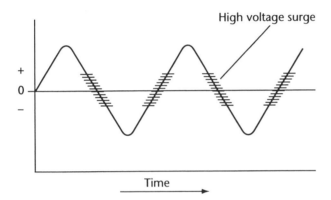

Figure 4.3   High voltage surge in AC supply.

Scratch start

With this method the tungsten is brought into contact with the work and gently scratched on the work surface to start the arc so that welding can begin. As mentioned above, contact between the electrode and the work is not the ideal situation and contamination may occur during the brief encounter. The contamination will normally be very small and in most cases will not seriously affect the strength of the weld. Having said that, this method should be restricted to less critical applications (see Fig. 4.4).

Lift start

With this method no high frequency is used, the electrode is brought into contact with the work at the start point of the weld, the torch switch or footpedal is pressed which makes contact, but no current flows in the circuit. At this point the electrode is drawn away from

the work creating an arc gap at which point the current flows, quickly rising (slope up) to the amperage set at the power source. When the switch is released the current gently reduces (slope down) until the arc is extinguished.

Slope up

This is a feature built into a more modern power source; it enables the operator to work out the time taken to reach the welding current chosen on the power source. This reduces the risk of burning away the plate edge at the start of the weld (especially on thin-sheet metals) and also contamination of the electrode tip. The time is set by means of a graduated switch (0–10), this being the time taken in seconds to reach the pre-selected current setting. This control is sometimes referred to as the 'soft start' control and can be switched off when not required.

Slope down

This is similar to slope up, but refers to the time taken to reduce from the welding current to no current. This gradual current reduction allows time for weld crater filling; the depression associated with the end of a weld run. The time is set using a graduated switch (0–20), this being the time taken in seconds to reduce from the welding current to the point where the arc is extinguished. This control is sometimes referred to as a crater filling device because of the function it carries out (see Fig. 4.5).

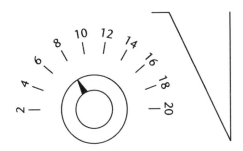

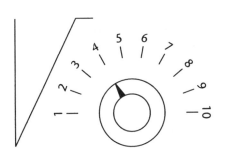

Figure 4.5 Slope up and slope down switches.

Footpedal

The footpedal can be used as an alternative to the slope up/slope down facility. It works on the principal that as the pedal is pressed the current increases (to the maximum set at the power source); alternatively, as the pedal is released the current will reduce to zero. The foot pedal has one additional feature in that the welding current can be regulated during the welding operation as well as at the start and finish of the weld.

Electrode types and diameters

Tungsten is a metal with a melting-point temperature of 3000°C which also possesses good electrical and thermal conductivity qualities. It was originally considered a suitable electrode material for the TAGS process. It was later discovered that by alloying the tungsten with 1–2% thorium or zirconium helped to improve arc stability, improve current carrying capacity, increase electrode life, resist contamination and improve arc striking. This

resulted in two types of electrodes being developed for manual TAGS welding: thorated electrodes for DC welding and zirconated electrodes for AC welding.

The electrode diameter is determined by the current and polarity. Recommended diameters are given in Table 4.2.

Table 4.2 Recommended electrode types for TAGS welding (courtesy of BOC)

Electrode type	Use
1–2% thoriated tungsten	DC electrode negative. Ferrous metals, copper, nickel, titanium
Ceriated tungsten	As above. Improved restriking and shape retention
Zirconiated	AC. Aluminium and magnesium alloys

Recommended Current Rating BS3019, Part 1		
	Maximum current amps	
	DC electrode (–)	AC electrode
Diameter (mm)	thoriated	zirconiated
0.8	45	–
1.2	70	40
1.6	145	55
2.4	240	90
3.2	380	150
4.0	440	210
4.8	500	275
5.6	–	320
6.4	–	370

Electrode grinding

The angle to which an electrode is ground depends on its application. The included angle or vertex angle (shown in Fig. 4.6) is usually smaller for low-current DC applications. To obtain consistent performance on a particular joint it is important that the same vertex angle is used.

To achieve consistency in the grinding of electrodes, grinding machines are available with a jig to hold the different sizes of tungsten electrodes. This jig ensures that regrind-

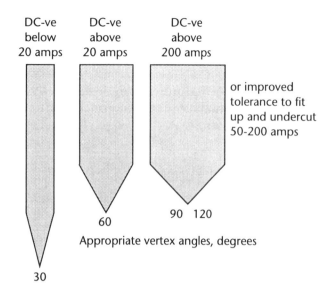

Figure 4.6   Preparation of GTAW electrodes showing approximate vertex angles and degrees. Note that electrodes for AC are normally used with a spherical tip (courtesy of BOC).

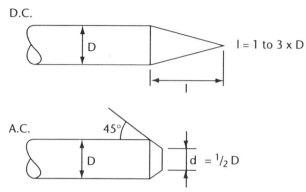

Figure 4.7   Ideal electrode shape for (a) DC welding and (b) for AC welding.

ing the electrode is done correctly and safely.

When freehand grinding of electrodes has to be carried out, care must be taken to maintain the point angle as this helps to concentrate the arc at the electrode tip, improve arc stability and operator control.

When using DC the electrode is ground to a point length of between one and three times the diameter of the electrode. The lower the current the longer the point. Electrodes are normally supplied in standard lengths of 75 and 150 mm and diameters 0.8, 1.2, 1.6, 2.4, 3.2, 4.0 and 6.4 mm.

When using AC a 45° chamber can be ground on one end of the electrode; this tends to help with the formation of the spherical shape required (balling up) usually carried out on a carbon block or a piece of the scrap metal being welded (see Fig. 4.7).

Collets

The electrode is held in the torch by a split copper tube known as a collet. The collet grips the electrode in the collet holder by tightening the electrode end cap. Collets are available in sizes to suit the various electrode diameters.

Gas nozzles

Ceramic nozzles are used mainly because of the resistance to heat generated during welding, although for currents above 200 amps water-cooled metal nozzles are available. Nozzles can be obtained in a variety of bore sizes ranging from 6 to 15 mm. The size of nozzle should be chosen to give correct gas protection to the weld pool, electrode, filler and, in some instances, the deposited metal. Various shapes are available to suit the job in hand, extended nozzles for welding in deep 'V', short nozzles for welding in confined spaces, transparent nozzles providing good visibility where the electrode projection (stick out beyond nozzle) must be kept to a minimum and gas coverage to a maximum.

Gas nozzles are designed to screw onto the collet holder, held by the torch body. The gas nozzle is a delicate piece of equipment; consistent heating and cooling can cause cracking and, in extreme cases, pieces break away making the nozzle unusable.

The collet holder screws onto the main body of the torch providing a means of clamping the collet and supporting the gas nozzle. The collet holder comes supplied with or without a gas lens.

Gas lens

Turbulence in the gas flow as it comes from the nozzle can give poor gas shielding of the weld area. This can be improved by using a gas lens, which will concentrate the gas over the weld area. Although the current carrying capacity is reduced with increasing electrode extension, using gas lenses allows the electrode extension to be increased, improving visibility and access in difficult situations. The use of the gas lens allows a wider variation in torch angle (slope), but, when combined with increased electrode extension, slope angles must be maintained near 90° to plate surface.

Torches

Welding torches for use with the TAGS process are either air or water cooled, fully insulated and are generally available with a flexible conduit assembly containing the power cable, gas and water tubes (if used). Some modern torches incorporate the switching device for starting the arc. The physical size of the torch is usually determined by the current capacity. Torches for use up to 200 amps are lightweight, small in size and available with flexible heads. The torch body supports the collet holder and, in turn, the collet and gas nozzle. The electrode end cap also screws into the torch body (see Fig. 4.8).

Shielding gas

Pure argon is the most popular choice of shielding gas supplied in a cylinder containing 11.01 m^3 at 200–230 bar pressure and painted light blue, or in bulk on-site storage tanks. To maintain quality the cylinder pressure should not be allowed to drop below 2 bar. Argon is reduced to a suitable working pressure by means of a single-stage regulator to control the gas supply, and a flowmeter is used to measure the quantity of gas being used. Argon gas is heavier than air and will displace oxygen; good ventilation should be ensured at all times. If this gas has to be used in confined spaces, adequate ventilation and correctly sited extraction equipment are essential to avoid any accumulation of gas. The use of oxygen meters is recommended to warn of low atmospheric oxygen levels. General safety precautions apply to the handling, storage and transportation of gas cylinders.

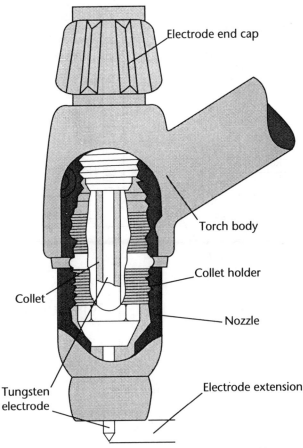

Figure 4.8 Torch assembly: torch body, collet holder, collet, tungsten electrode, electrode end cap and ceramic.

Regulator and flowmeter

A typical gas flow for TAGS welding is 8–12 litres/min. The regulator and flowmeter are shown in Fig. 4.9.

A second flowmeter can be connected to the regulator and used to supply and regulate gas flow to the root side of the joint (gas backing). For example, a gas backing is required when welding stainless steels, especially for the initial runs. The gas is piped to the root side of the joint excluding air and preventing atmospheric contamination. Available gas mixtures and their applications are listed in Table 4.3.

AC/DC non-consumable electrode arcs

When using DC approximately two-thirds of the total heat is concentrated at the positive

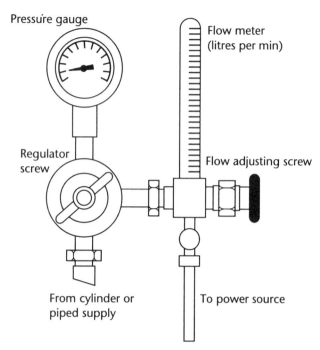

Typical gas flow for TAGS welding is 8-12 litres per minute

Figure 4.9 Regulator and flowmeter.

Table 4.3 Available gas mixtures and their applications (courtesy of BOC)

Gas	Applications	Features
Argon	TAGS all metals	Stable arc performance. Efficient shielding. Low cost
Helium	TAGS all metals especially copper and aluminium	High heat input. Increased arc voltage
Argon + 25–80% helium	TAGS aluminium, copper, stainless steel	Compromise between pure argon and pure helium. Lower helium contents normally used for GTAW
Argon + 0.5–15% hydrogen	TAGS austenitic some copper stainless steel, nickel alloys	Improved heat input, edge wetting and weld bead profile

pole and one-third at the negative pole. Being able to choose between electrode positive and electrode negative enables the choice of heat at the electrode or the workpiece.

Direct current electrode negative (DC–), when the electrons flow is from the electrode to the workpiece means that one-third of the heat is concentrated at the electrode and two-thirds at the workpiece. The total result is a narrow weld bead with good penetration.

Direct current electrode positive (DC+), when the electrons flow from the workpiece to the electrode means that two-thirds of the heat is concentrated at the electrode and one-third at the workpiece. In this situation the electrode becomes overheated; possibly melting to become rounded in shape. The total result is a wide weld bead with little penetration. However, using DC+ electrode the removal of surface oxides from metals is exceptionally good: in welding metals such as aluminium for example. The oxide removal only takes place when the electron flow is away from the weld pool, but as already mentioned this is accompanied by overheating of the electrode and low penetration.

AC creates a situation where it is possible to change between electrode positive and electrode negative. These changes take place at 100 times per second, which results in the total heat available being divided equally between electrode and workpiece. When the electrode is positive the arc action will lift any oxides from the surface of the work; the weld pool appearing bright and clean. When the electrode changes to negative the electrode is cooled and the penetration achieved is between that when using electrode positive or negative (see Figs 4.10 and 4.11).

Filler rods

The filler rods used in TAGS welding conform to BS2901 Part 1, Filler Rods and Wires for Gas Shielded Arc Welding of Ferritic Steels, which covers the chemical composition, diameter and tolerance of rods and wires, condition of rods and wires, dimension of reels of wire, packaging and marking. Fillers are available in 1.6, 2.4 and 3.2 mm diameter × 750 mm long and copper-coated for steels designed specially

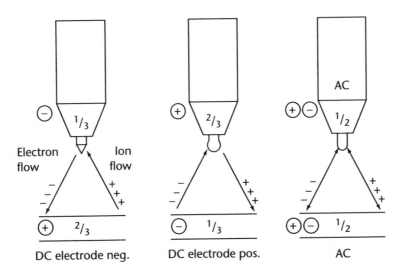

Figure 4.10 Arc heat distribution.

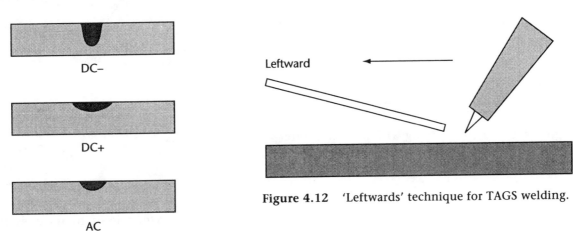

Figure 4.11 Effect of current flow on penetration.

Figure 4.12 'Leftwards' technique for TAGS welding.

for TAGS welding. Filler rods should be cleaned with wire wool before welding. Care should also be taken to avoid picking up grease or moisture by handling. Thin leather or fireproof cotton gloves are recommended when welding, as filler wire manipulation and control is essential in the achievement of quality welds.

Direction of travel

TAGS welding can be carried out successfully in all positions using the 'leftwards' technique, where the filler precedes the torch in the direction of travel. In the case of a right-handed person the torch is held in the right hand, filler in the left hand, travelling from right to left (see Fig. 4.12).

Stops and starts

The requirements for stops and starts are the same as previously discussed, e.g. consistent in width with no abrupt change in shape. To restart the weld the electrode is positioned at the back edge of the previous weld crater. When the arc is established the torch is moved slightly forward into the original weld pool and held until the full molten weld pool is formed. At this point, filler metal can be added to the weld pool and progression made along the joint. Care must be taken to avoid the filler coming into contact with the electrode.

Fusion runs

Welds can be made without the use of a filler (fusion runs), but some joint types require preforming to accommodate the weld. To perform acceptable fusion welds the joint edges must be tightly fitted up and tacks will be closer together. This technique is limited to steels up to 1.6 mm thick. Some typical joint designs are shown in Fig. 4.13.

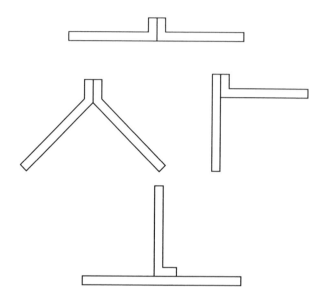

Figure 4.13 TAGS welding without filler; different arrangements for fusion runs.

Weaving

Weave patterns have been discussed extensively in the two previous chapters on MMA and MAGS: the same techniques apply with the TAGS process. The main difference being that with the TAGS process weave patterns will relate to torch manipulation with the application of filler metal applied separately from a filler rod. This will also apply to oxyacetylene welding. Weave patterns and filler rod applications will be dealt with in TAGS positional welding.

Process applications

The use of this process finds its way into many industries, particularly nuclear, aircraft, chemical plant and equipment, automotive, brewing and food processing. This is mainly due to the process eliminating fluxes and not leaving corrosive residues behind after welding. In the hands of a skilled operator the TAGS process

will produce quality welds on carbon and stainless steels and is used for welding sheet metal and thin wall (up to 8 mm) pipe welds. It is also used for depositing root runs in thick wall pipes to achieve control and quality of root penetration. The success of the process in depositing root runs led to the introduction of the fusible root insert, a pre-formed metal insert which is tack welded into the root when setting up the joint. By using a fusion run technique (without filler) the root insert is fused with the root faces of the joint (see Fig. 4.14).

Fusible root insert

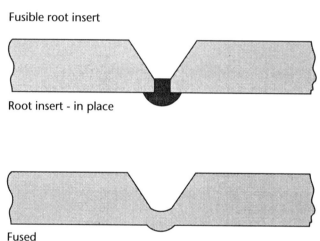

Root insert - in place

Fused

Figure 4.14 Good joint fit up is essential when using root inserts.

The process is suited to producing quality welds on aluminium and its alloys, magnesium and its alloys, copper, nickel alloys and titanium. It can be used for depositing wearsurfacing materials for protection against abrasion, impact and corrosion. The process also lends itself readily to automation.

TAGS positional welding

If you have progressed through the text in a logical order, you should have sufficient knowledge on the weave patterns used and the electrode angles associated with the different positions. As mentioned earlier in this chapter, TAGS and oxyacetylene welding differ slightly in that both torch and filler rod angles have to be considered. These will be explained relative to this process and positions in this chapter. If on the other hand this is your start point it is advisable to read the sections on weaving in the MMA and MAGS chapters.

It is not the intention of this book to cover all the welding positions, but to give a guide to the torch and filler rod angles, sizes of electrodes and amperages and techniques used for a range of welding positions. Slight variations may be required depending on condition of equipment, thickness and type of material, joint type, preparation and cleanliness of material.

The action of dipping the filler wire into the weld pool will have a slight cooling effect; therefore, care should be taken to avoid chilling the weld pool, which may result in lack of fusion or penetration and, in extreme cases, freezing the filler into the weld pool.

Care should also be taken to ensure that when the dipping action of the filler rod takes place the end of the filler is not withdrawn from the protection of the gas shield as this may introduce impurities into the weld.

Although sizes of filler are recommended below, the influencing factors governing the choice of filler will be:

a) plate thickness
b) joint type
c) joint preparation
d) current setting
e) welding position.

Care should also be taken to ensure that when movement of the filler takes place the end is not withdrawn from the protection of the gas shield as this may introduce impurities into the weld.

Horizontal–vertical 'T' fillet

See Table 4.4. Using the leftward technique, commence the weld at the right-hand side of the joint by forming the weld pool equal on both horizontal and vertical plates. When the weld pool is established progress the weld along the joint, feeding the filler into the weld pool with regular, consistent movements. The filler rod should be fed into the leading edge of the pool and slightly toward the vertical

plate. Spot tacks should be sufficient to restrain plates during welding and approximately 50–75 mm spacing minimum weaving will be required (see Fig. 4.15).

Table 4.4 Horizontal vertical 'T' fillet on a 1.6 mm thick plate welded in the flat position, using a 1.6 mm filler rod and a 1.6 mm electrode DC– at 70–95 amps

	Slope angle	Tilt angle
Torch	70–80°	Equal to either side
Filler	10–20°	Equal to either side

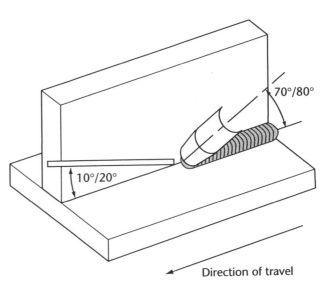

Figure 4.15 Torch and filler rod angles.

Horizontal–vertical corner joint

Horizontal vertical corner joint on 3 mm thick plate welded in the flat position using a 2.4 mm filler rod and a 2.4 mm electrode DC at 90–140 amps. Slope and tilt angles for the torch and filler rod will be as in Table 4.4, using the same procedure as for the horizontal vertical fillet, but concentrating the filler rod more toward the top of the vertical edge of the joint. Although it is not always necessary, by maintaining the same slope angle and feeding the filler wire into the weld pool on the top edge at 90° to the joint line will help prevent excessive sagging of the molten weld metal which may result in overlapping. Tack at each end at 75–100 mm spacing. Weaving should not be required (see Fig. 4.16).

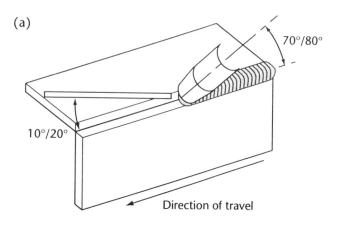

(a)

70°/80°

10°/20°

Direction of travel

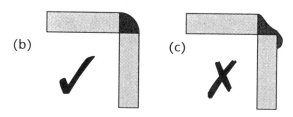

(b) ✓   (c) ✗

Figure 4.16   Outside corner joint: (a) torch and filler rod angles, (b) correct profile, (c) incorrect profile.

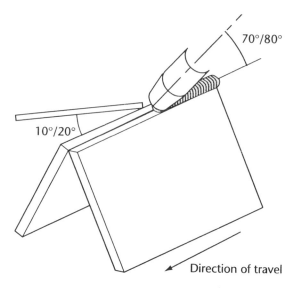

70°/80°

10°/20°

Direction of travel

Figure 4.17   Outside corner: torch and filler rod angles.

Outside corner joint

See Table 4.5. After tacking establish the arc at the right-hand end of the joint, allowing the weld pool to form until it melts the uppermost edges of the joint. The filler rod is then applied to the leading edge of the weld pool with a regular, consistent dipping action. Progress the weld along the joint at a speed consistent with the formation of the weld pool. Spot tack at intervals of 50–75 mm (see Fig. 4.17).

Square edge butt joint

See Table 4.6. Although the joint is a square edge, a slight chamber can be filled or ground on the face side. This will aid visibility of the joint line without adverse effect on the joint. As the weld pool is formed a slight sinking will occur. The filler rod should be applied to the pool at this point and progression made along the joint. Tack welds should be of sufficient strength to restrain the joint during welding at 75–100 mm spacing. Pre-setting may be necessary if a jig is not used (see Fig. 4.18).

Table 4.5   Outside corner joint on 1.6 mm plate welded in the flat position using a 1.6 mm filler rod and a 1.6 mm electrode DC– at 60–90 amps

	Slope angle	Tilt angle
Torch	70–80°	90 ± 5°
Filler rod	10–20°	Equal to either side

Table 4.6   Square edge butt joint on 3 mm thick plate welded in the flat position using a 2.4 mm filler rod and a 2.4 mm electrode DC– at 90–180 amps. A slight root gap may be required (1–2.5 mm)

	Slope angle	Tilt angle
Torch	70–80°	90 ± 5°
Filler rod	5–15°	Equal to either side

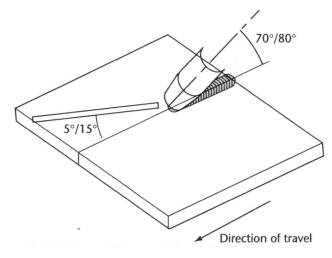

Figure 4.18   Torch and filler rod angles.

Horizontal–vertical single 'V' butt joint

See Table 4.7. The weld preparation for this position is 60–70° included angle with 15–20° angle on the bottom edge of the joint. A slight root face (1 mm) may be an advantage on the top plate no gap. Establish the arc at the right-

Table 4.7   Single 'V' butt joint on a 4 mm thick plate welded in the horizontal–vertical position using a 3.2 mm filler rod and a 2.4 mm electrode DC– at 100–200 amps

	Slope angle	Tilt angle
Torch	70–80°	90 ± 10°
Filler rod	10–20°	90 ± 10°

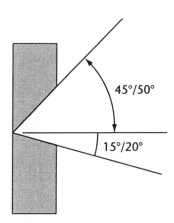

Figure 4.19a   Joint preparation.

hand end of the joint to form the weld pool. This will be to a greater extent on the upper plate. When the pool is established, a filler rod can be added, dipping it in the top edge of the weld pool and allowing it to fuse into both plate edges. For subsequent runs, torch and filler rod angles should be adjusted within the tolerances to achieve the correct weld profile (see Figs 4.19 and 4.20).

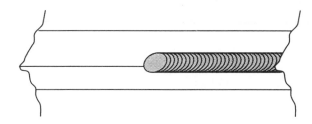

Figure 4.19b   Weld pool formation.

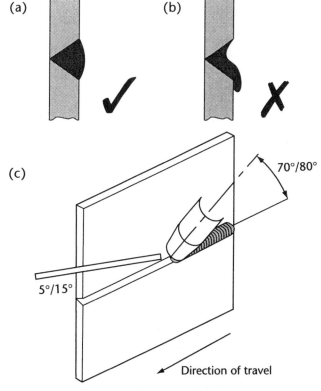

Figure 4.20   (a) Correct profile, (b) incorrect profile, (c) torch and filler angles.

Vertical fillet weld

See Table 4.8. After tacking up the joint, establish the arc at the bottom of the joint and hold until the weld pool is formed equally on each plate, either side of the root. At this point the filler rod can be introduced from the top into the top edge of the weld pool. Small amounts of filler metal should be added with a regular consistent dipping action. Progression along the joint will be consistent with the formation of the weld pool. Weaving will be the minimum to achieve weld shape, size and fusion at both sides of the weld (see Fig. 4.21).

Table 4.8 Vertical fillet weld on a 3 mm thick plate welded vertical up using a 2.4 mm filler rod and a 2.4 mm electrode DC− at 100–200 amps

	Slope angle	Tilt angle
Torch	80–90°	Equal to either side
Filler rod	5–20°	Equal to either side

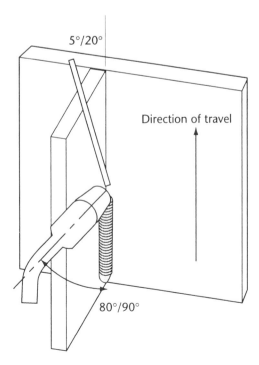

Figure 4.21 Torch and filler rod angles.

Vertical corner joint

See Table 4.9. Tack weld the two plates together ensuring the two inner edges of the joint are aligned. Establish the arc at the bottom of the joint and form the weld pool equally on each plate edge either side of the root. For the root deposit, pay particular attention to the melting of the two root edges. A small hole (weld onion) may appear to indicate this. At this point the filler wire is introduced from the top to the leading edge of the weld pool. Speed of travel will be consistent with the melting of the root edges. On completion of the root deposit establish the arc and weld pool at the bottom of the joint, this time ensuring that melting takes place over the root run and the outer edges of the joint. This may require slight side to side weaving of the torch. Filler metal can be added to the weld pool at this stage, which will flow across the joint with the weaving motion of the torch. A slight pause at the edges of the joint will ensure good fusion. Care must be taken to avoid adding excess filler which will cause sagging of the molten weld metal, resulting in an incorrect weld profile (see Fig. 4.22).

Single 'V' butt joint

See Table 4.10. After tacking start welding the root run by establishing the arc at the lower end of the joint, form the weld pool to achieve melting on both edges of the joint. At this stage apply the filler rod from the top into the leading edge of the weld pool, progressing up the joint adding filler at the desired rate, avoiding chilling of the weld pool and freezing the filler into the pool. For the capping run, establish the arc at the lower end of the joint and form the weld pool, ensuring melting of

Table 4.9 Vertical corner joint on a 5 mm thick plate, welded vertical up using a 2.4 mm filler rod and a 3.2 mm electrode DC− at 100–200 amps

	Slope angle	Tilt angle
Torch	70–90°	Equal to either side
Filler rod	10–20°	Equal to either side

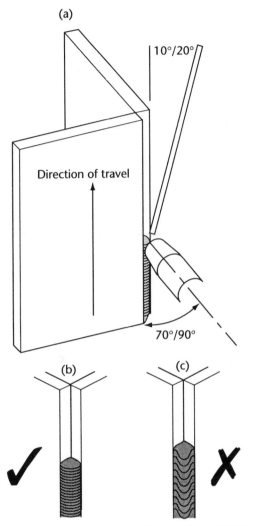

Figure 4.22   Outside corner: (a) torch and filler rod angles, (b) correct profile, (c) incorrect profile.

the root run and the outer edges of the joint by a side to side weave of the torch. The filler wire can be kept on centre with no movement, providing sufficient filling of the joint is achieved. Alternatively the filler metal can be dipped into the weld pool at the extremities of the weave; the molten metal will form the weld bead shape with the manipulation of the torch (see Fig. 4.23).

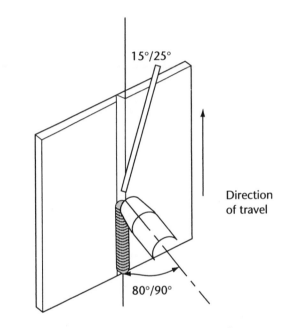

Figure 4.23   Torch and filler rod angles.

Table 4.10   Single 'V' butt joint on a 5 mm thick plate, welded in the vertical position using a 2.4 mm filler wire and a 2.4 mm electrode DC– at 100–200 amps, with an included angle of 70–90°. A root face of 1.0–2.0 mm and a root gap of 0–1.6 mm can be used to achieve the required degree of penetration

	Slope angle	Tilt angle
Torch	80–90°	Equal to either side
Filler rod	15–25°	90 ± 10°

5

Oxyacetylene welding and cutting

Oxyacetylene welding (sometimes referred to as gas welding) gets its name from the two gases commonly used in the process: oxygen and acetylene. This process can be used for welding a wide variety of ferrous and non-ferrous metals, the latter requiring a flux in most cases. This process of joining finds a place in both repair work in the motor vehicle body repair trades, to the fabrication of new components, such as ductwork and pipework installations for heating, ventilating and sprinkler systems.

Standard welding/cutting equipment

The standard welding/cutting equipment is shown in Fig. 5.1.

The high-pressure system uses acetylene gas supplied in cylinders. It is the most common form of acetylene supply and one which is most widely used by industry. The nominal content of an acetylene cylinder is 8.69 m^3. The acetylene gas is dissolved in a solvent, acetone, and is supported in a porous mass of Kapoc. Acetylene is a colourless gas, which is extremely flammable, with a garlic-like odour and a flammability range of 2.4–88% volume in air. Metal alloys containing more than 70% copper or 43% silver must not be used in acetylene systems as this can result in the formation of dangerous or explosive components. The acetylene cylinder has two identifying features:

■ its colour: acetylene cylinders are painted maroon, and

■ fittings: the cylinder valve has a left-hand thread. This is common to all fittings associated with acetylene.

The other gas used in the process is oxygen (O$_2$) supplied in cylinders up to capacities of 9.66 m^3 and 200 bar pressure. Oxygen is colourless, odourless and although non-flammable it strongly supports combustion. Care should be taken to avoid clothing or other combustible materials becoming saturated with the gas. The oxygen cylinder has two identifying features:

■ colour: oxygen cylinders are painted black, and

■ fittings: the cylinder valve has a right-hand thread. This is common to all fittings associated with oxygen.

Safety precautions to be observed when using oxygen and acetylene cylinders

a) Always keep cylinders, fittings and connections free from oil and grease.
b) Store cylinders in the upright position.
c) Transport cylinders in the upright position.
d) Store flammable and non-flammable gases separately.
e) Store full and empty cylinders separately.
f) Avoid cylinders coming into contact with heat.
g) Secure cylinders by 'chaining up' during storage, transport and use.

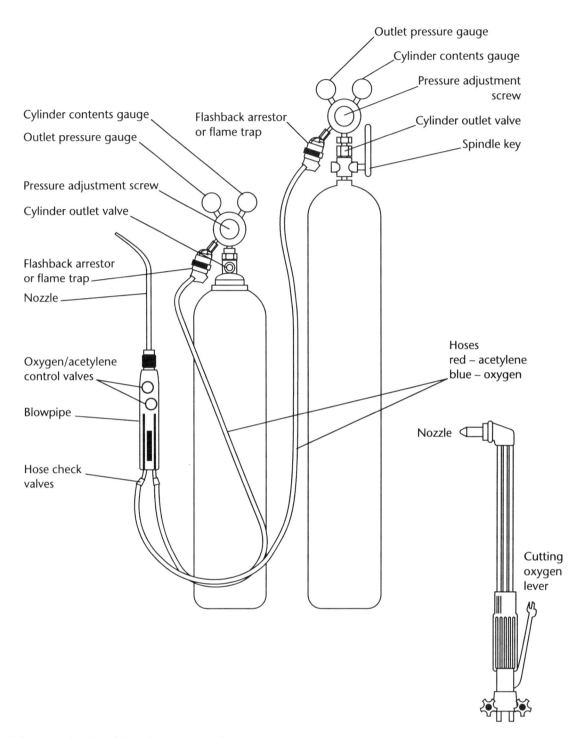

Figure 5.1 Standard welding/cutting equipment.

h) Open cylinder valves slowly.
i) Only use a recommended leak-detection spray.
j) Report damaged cylinders to the supplier.
k) Always check for leaks.
l) Switch off gas supply after use.

A fuel gas widely used with oxygen is Apachi (trade name) which is classified as a liquid petroleum gas (LPG). This gas is available commercially in portable steel containers or bulk quantities. It is a highly flammable gas which can cause severe cold burns to the skin and damage to the eyes. The gas is widely used for cutting, brazing, soldering, metal spraying and pre-heating but is not used for welding. These applications are normally safe because the burning of the fuel gas and oxygen is done in a controlled manner using purpose-designed equipment. Users of this type of equipment should be aware of the possibility of fuel gas feeding back into the oxygen line and vice-versa when the equipment is not in use. The explosive gas mixture formed can detonate violently when the equipment is re-ignited. Safe use can be ensured by:

■ observing correct operating procedure
■ correct purging
■ use of flashback arrestors
■ use of hose check valves
■ use of suitable nozzles.

Only equipment designed for the use with this gas should be used. Sources of leaks should be checked with a recommended leak-detection spray. *Only hoses and regulators specifically designed for Apachi gas should be used. Alloys with a copper content of over 60% are not suitable for use in these systems.*

Gas cylinders are colour-coded: lilac cylinder with a red valve guard and shoulder; fittings are left-hand threaded. The gas should be stored and used with adequate ventilation. It is heavier than air and in the event of a leak will spread and accumulate in low-lying areas where it is slow to disperse.

Regulators

The pressure regulator used with the oxyacetylene process is a precise measuring instrument and should be treated and handled with care. It consists of a pressure-regulating screw for adjusting working pressure, and two gauges: one indicates the cylinder contents pressure and the other the working pressure set at the regulator. The two regulators are shown in Fig. 5.2.

Included in the safety features built into the regulator is a safety valve which will discharge excess pressure very quickly, resealing when correct pressure is restored. Fittings on the regulator are left-hand threaded for acetylene and right-hand threaded for oxygen. All the connection nuts on the acetylene regulator will

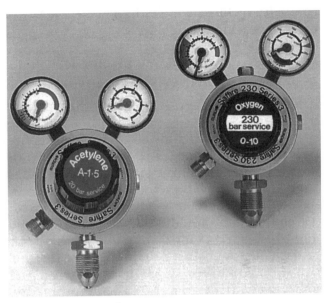

Figure 5.2 Regulators.

have grooved corners. This is common to all acetylene connections. The regulator is also identified by colour: red for acetylene and blue for oxygen. Regulators are available in single- and multi-stage versions. This relates to the stages in which the cylinder pressure is reduced to the operating pressure. Multi-stage regulators allow a more stable outlet pressure to be obtained. Regulators are manufactured for specific gases and uses. Using the correct type of regulator will help in preventing dangerous events. Never use sealing tapes or compounds on fittings. If in doubt seek advice from the manufacturer or supplier.

Safety

Two dangerous events associated with oxyacetylene welding or cutting are:

■ backfire and
■ flashback.

Backfire is the entry of the flame into the neck of the cutting or welding torch, which is usually extinguished at this point. This does not usually result in damage to the torch, although the cause must be identified and eliminated; the torch being allowed to cool before commencing. Possible causes are:

a) the nozzle being too close to the weld pool or plate surface

b) faulty or loose connections
c) overheating of the nozzle
d) incorrect gas pressures
e) a foreign body entering the nozzle
f) faulty or wrong equipment.

Flashback is a progression of the backfire where the flame passes past the torch and into the hoses which, if allowed to progress, could result in serious damage to the equipment, or possibly an explosion. In the event of a flashback the torch valves should be turned off, oxygen first, then both cylinder valves closed. All equipment should be dismantled, cleaned and inspected, and any defective items replaced before using. Comprehensive instructions are contained in the booklet *Safe Under Pressure* available from BOC. Some protection is afforded in the system by the use of hose protectors and flashback arrestors. These are discussed in more detail below.

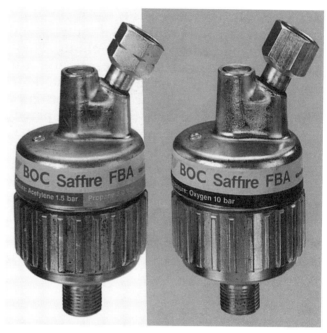

Figure 5.3 Flashback arrestors.

Flashback arrestors

The flashback arrestor (flame trap) is a safety device, highly sensitive to temperature and pressure, fitted to the outlet side of the oxygen and acetylene regulator (see Fig. 5.3). It is a very sensitive cut-off device. The arrestor will activate at the slightest indication of back pressure and prevents backfeeding to the regulator and cylinder, at the same time cutting off the gas flow. The flashback arrestor is effective in the event of a backfire by:

a) extinguishing the backfire
b) stopping the flow of gases after a backfire
c) preventing reverse pressure flow
d) protecting operator and equipment.

In the event of a backfire the flashback arrestor must not be re-set until the cause has been identified and removed.

Hoses (BS5120)

The hose is a means of conveying the gas to the welding or cutting torch. It is made of synthetic rubber with fabric reinforcement. Hoses are available in lengths of 5–20 metres and bore diameters of 6.3–10.00 mm and colour coded **red** for acetylene and **blue** for oxygen.

They are designed to be lightweight, oil resistant and to cope with normal working conditions. Following a few simple guidelines will ensure long life and safe operation of hoses:

1 Ensure hoses are free from burns, cuts and cracks.
2 Avoid dragging hoses over sharp edges and objects.
3 Do not wrap hoses around cylinders when in use or stored.
4 Avoid exposure to heat and hot objects.
5 Do not use hose longer than is necessary for the job.
6 Avoid contact with oil and grease.
7 Route hoses to avoid traffic pressures.
8 Keep hose clear of sparks from metal being cut.
9 When connecting two lengths of hose only use approved couplers.
10 Ensure hoses are fitted with the correct connections.
11 Ensure hoses conform to British Standards.
12 Always test for leaks with an approved leak-detection spray.
13 Use correct colour-coded hoses.
14 Do not bind hoses together.
15 Do not use damaged hoses.

Purging

Hoses should be purged (blown through) using oil-free compressed air. **Do not use oxygen.**

Hose connectors

For connecting the hoses to the torch at one end and the flashback arrestor at the other, two types of connections are available

- the nipple and nut type – hose to flashback arrestor
- the hose check valve – hose to welding/cutting torch incorporating a non-return valve.

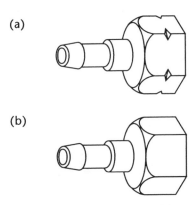

Figure 5.4 Nipple and nut type: (a) grooved hexagon nut for acetylene, (b) plain hexagon nut for oxygen.

The nipple and nut type (see Fig. 5.4) is a 'straight through' connector used to attach the hose to the flashback arrestor. The connections are available in size to suit the bore diameter of the hose. The hexagon nut of the acetylene connections is grooved and left-hand threaded while the oxygen connector has a plain hexagon nut with a right-hand thread.

Hose check valves

The hose check valve (see Fig. 5.5), or hose protector as it is more commonly called, is used to attach the hoses to the welding or cutting torch. It incorporates a non-return valve, and acts as a safety device to prevent the back-feeding of gases (premature mixing of the

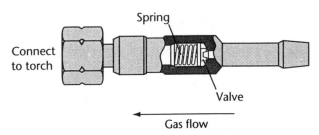

Figure 5.5 Hose protector (hose check valve).

gases in the hoses). Backfeeding is one of the main causes of a backfire; the fitting of hose protectors will prevent this reaching the risk of fire, damage to equipment and possible injury to the operator. The hexagonal nut of the acetylene connector is grooved and left-hand threaded, while the oxygen connector has a plain nut with a right-hand thread.

'O' clips

The 'O' clip (see Fig. 5.6) is a simple metal ring used for fastening the hoses to the hose connectors. The clip is placed over the hose, the connector is inserted into the hose and the 'O' clip is crimped onto the hose using a suitable pair of pincers. Care should be taken to avoid cutting into the metal of the 'O' clip, which will weaken it and cause possible failure in service.

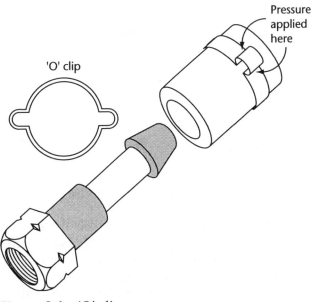

Figure 5.6 'O' clip.

Welding torches

The welding torch consists of a shank which incorporates the oxygen and acetylene flow control valve, a mixer and a range of copper nozzles. As mentioned previously, two types are available:

■ high pressure and
■ low pressure.

The high-pressure torch

The high-pressure torch (see Fig. 5.7) is designed for use with high-pressure gases, supplied from cylinders. It acts as a mixing device when supplying approximately equal pressures of oxygen and acetylene where they are mixed prior to being burnt at the nozzle tip. Using the high-pressure system one mixer can accommodate different nozzle sizes to enable the welding of a wide range of metal thicknesses, from 1 to 25 mm. The mixer system ensures that there is a least amount of explosive mixed gas in the system, prior to being burnt. In general, the welding process is better performed with mixer-type torches. High-pressure torches **must not** be used with low-pressure systems.

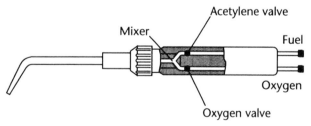

Figure 5.7   High-pressure torch.

The low-pressure torch

The low-pressure torch uses the injector system (see Fig. 5.8) where the higher pressure oxygen draws the lower pressure acetylene into the mixing chamber. The injector is usually included as part of the nozzle, so in this respect each nozzle has its own injector. This type of nozzle is far more expensive than those used with the high-pressure system. The injector principle requires that the oxygen pressure

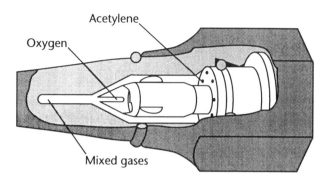

Figure 5.8   Injector. As the oxygen issues at relatively high velocity from the tip of the injector, it draws the proper amount of acetylene into the stream. The oxygen and acetylene are thoroughly mixed before issuing from the blowpipe tip.

is always higher than the acetylene pressure. This means that as the nozzle becomes blocked the oxygen can backfeed into the acetylene line causing premature mixing of the gases and increasing the risk of backfires. The low-pressure torch is shown in Fig. 5.9.

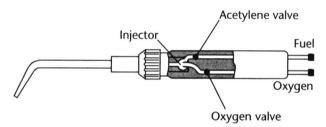

Figure 5.9   Low-pressure torch.

Nozzles

The heat value to the flame is governed by the size of the hole in the nozzle being used. The heat value of the flame will need regulating to suit the thickness, thermal conductivity, joint type and melting temperature of metal being welded. This should be achieved by changing the nozzle size and not necessarily changing the pressure. For example, metal up to 3 mm thick can be welded by changing the nozzle size without excessive alteration to gas pressure (see Table 5.1).

A flame with too low a heat value will result in lack of penetration and fusion. Backfiring may also occur. If the flame heat is too high, overheating, lack of control of the molten metal and **weld** pool and excessive penetration

Table 5.1 Nozzle sizes and material thicknesses (courtesy of BOC)

| | Mild steel | Tk'ness | Nozzle | Operating pressure | | | | Gas consumption | | | |
| | | | | Acetylene | | Oxygen | | Acetylene | | Oxygen | |
mm	in	swg	size	bar	lbt/in^2	bar	lbt/n^2	l/h	ft^3/h	l/h	ft^3/h
0.9		20	1	0.14	2	0.14	2	28	1	28	1
1.2		18	2	0.14	2	0.14	2	57	1	57	2
2		14	3	0.14	2	0.14	2	86	3	66	3
2.6		12	5	0.14	2	0.14	2	140	5	140	5
3.2	1/8	10	7	0.14	2	0.14	2	200	7	200	7
4	3/32	8	10	0.21	3	0.21	3	280	10	280	10
5	3/16	6	13	0.28	4	0.28	4	370	13	370	13
6.5	1/4	3	18	0.28	4	0.28	4	520	18	520	18
8.2	8/16	0	25	0.42	6	0.42	6	710	25	710	25
10	3/8	4/0	35	0.83	9	0.63	9	1000	35	1000	35
13	1/2	7/0	45	0.35	6	0.35	6	1300	45	1300	45
25	1+		90	0.83	9	0.83	9	2500	90	2500	90

will be experienced. Effective flame shape and heat values can only be maintained if the nozzle hole is clean and square with the end of the nozzle. This should be carried out by using a recommended nozzle reamer (nozzle cleaner), inserting the reamer into the hole and, with a gentle twisting motion, removing any obstruction (see Fig. 5.10).

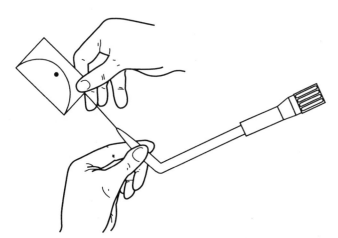

Figure 5.10 Cleaning the nozzle hole.

Equipment assembly (welding and cutting)

Secure cylinders to a suitable trolley or stand by chaining or clamping to prevent them from falling or being pulled over. Failure to do so will allow acetone to be drawn from the acetylene cylinder.

Before connecting the regulator to the cylinder 'snifting' (briefly opening and closing the cylinder outlet valve) is recommended to remove any dust or contamination from the outlet. Cylinders are opened and closed by either a valve wheel or by the use of a spindle key designed for this purpose (see Fig. 5.11). During use the spindle key should be left in the acetylene cylinder outlet valve (if only one key is available). It is good practice to attach the key to the cylinder trolley or stand. Connect the respective regulator to the cylinder checking that the seating face is not damaged. Right-hand threads = oxygen; left-hand threads = acetylene. The regulator should be sited so that it does not interfere with the opening and closing of the cylinder valve, which should be possible in a single-movement operation.

Fit flashback arrestors to both oxygen and acetylene cylinders, use the correct size of spanner and avoid over-tightening. After purging the hoses fit the nipple and nut fitting to the flashback arrestor; avoid over-tightening and damage to connecting nuts. Connect the other end of the hoses (hose protector) to the respective fitting on the welding/cutting torch. Do not over-tighten. Select and fit a suitable welding/cutting nozzle for the job in hand. This completes the assembly of the equipment.

Before opening the oxygen and acetylene cylinder valves, ensure that the pressure adjustment screws on each regulator are released (turned fully anticlockwise). These

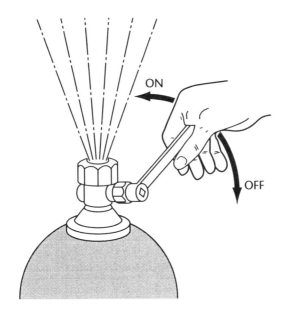

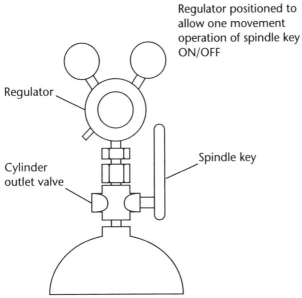

Figure 5.11 Snifting.

should be 'captivated' so that it cannot be screwed completely out of the regulator. The cylinder valves can then be slowly opened, the working pressure is adjusted to suit the work in hand by turning the pressure adjuster clockwise. The oxygen and acetylene valves on the torch can be opened (one at a time) and final adjustments made to the working pressure.

Lighting-up procedure

Open the acetylene control valve on the torch,

allowing the gas to flow for 5–10 s and light the gas with a suitable spark lighter. Keep the spark lighter and your hands out of line of the flame.

The acetylene flame should burn without smoke or sooty deposit. Adjust the flame if necessary to the condition.

Open the oxygen control valve on the torch and adjust until the inner core is clearly defined.

Flame settings

By adjusting the quantities of oxygen or acetylene fed to the flame, it is possible to achieve three types of flame condition: neutral flame, oxidising flame and carburising flame.

Neutral flame

The neutral flame (see Fig. 5.12) consists of approximately equal quantities of oxygen and acetylene being burnt. The flame has a clearly defined central (inner) cone with a length two to three times its width. This flame is used for most welding applications on steel, cast iron, copper and aluminium. This is the hottest flame condition reaching 3000–3200°C at a point approximately 3 mm forward of the inner cone.

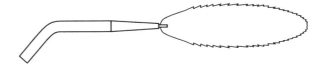

Figure 5.12 Neutral flame.

Oxidising flame

The oxidising flame (see Fig. 5.13) consists of an excess quantity of oxygen being burnt in the flame. It is obtained by first setting the neutral flame and then slightly increasing the quantity of oxygen. The inner cone will take on a much sharper pointed shape and the flame is much more fierce with a slight hissing sound. This flame type is used for welding brasses. It should be avoided when welding steels.

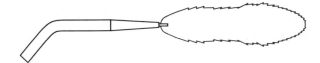

Figure 5.13   Oxidising flame.

Welding techniques

The process takes advantage of two techniques: 'leftward' (filler rod precedes the torch) and 'rightward' (torch precedes filler rod), in the direction of welding (see Fig. 5.15).

Carburising flame

The carburising flame (see Fig. 5.14) consists of an excess quantity of acetylene burning in the flame. Again, the best way of achieving this condition is to start with the neutral flame and then slightly increase the amount of acetylene. The inner cone will become surrounded by a 'feather' as a result of the excess acetylene. This flame is used for depositing of hard surfacing materials; it should be avoided when welding steels.

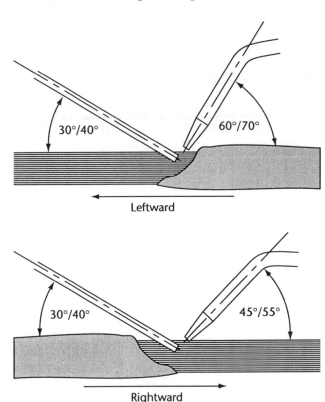

Figure 5.15   Leftward and rightward techniques.

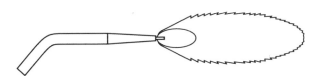

Figure 5.14   Carburising flame.

Extinguishing the flame

When welding has finished or is paused the flame is extinguished by first closing the acetylene valve at the torch shutting off the gas supply immediately, followed by closing the oxygen valve on the torch. Place the torch in a safe position where it cannot be damaged.

If the equipment is to be left for any length of time, leave it in a safe condition by:

1 Closing the cylinder valves on the acetylene then oxygen cylinder.
2 Opening and closing the acetylene valve on the torch.
3 Opening and closing the oxygen valve on the torch. Note that this will release the pressure from the hoses and the working pressure gauge on the regulator should register zero.
4 Releasing the pressure adjustment screw on the regulator (turn anticlockwise).

The rightward technique enables the welding of thicker sections (over 6 mm) but the economics of the process would probably dictate the choice of a more suitable means of joining metals above this thickness, e.g. MMA, MAGS.

High- and low-pressure systems

The process can operate from a high- or low-pressure system which gets the name from the method of acetylene supply. The low-pressure system uses acetylene produced locally to the point of use, and requires the use of a special low-pressure torch. The high-pressure system uses acetylene supplied in cylinders.

Filter glasses

Use the following filters when oxyacetylene welding:

3 GWF for aluminium and alloys
4 GWF for brazing and bronzing welding
5 GWF for copper and alloys
6 GWF for thick plate and pipe.

Plate edge preparation

Figure 5.16 shows the plate edge preparation depending on the thickness of the material to be welded.

Oxyacetylene positional welding

As mentioned in the chapter on TAGS welding, both the filler rod and torch angles have to be considered when using the oxyacetylene welding process. A selection of welding positions will be covered in this section giving torch and filler rod angles, size of filler rod and nozzle size(s).

The information relating to oxyacetylene positional welding is intended as a guide only. Slight variations may be required depending on thickness and type of material, joint type, preparation and welding techniques (leftward or rightward).

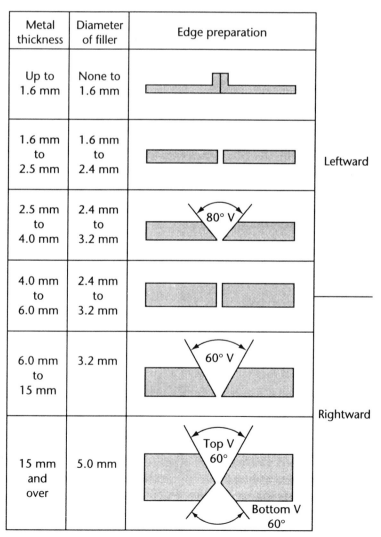

Metal thickness	Diameter of filler	Edge preparation
Up to 1.6 mm	None to 1.6 mm	
1.6 mm to 2.5 mm	1.6 mm to 2.4 mm	Leftward
2.5 mm to 4.0 mm	2.4 mm to 3.2 mm	80° V
4.0 mm to 6.0 mm	2.4 mm to 3.2 mm	
6.0 mm to 15 mm	3.2 mm	60° V — Rightward
15 mm and over	5.0 mm	Top V 60° / Bottom V 60°

Figure 5.16   Plate edge preparation.

The action of dipping the filler rod into the weld pool will have a slight cooling effect; therefore care should be taken to avoid chilling the weld pool which may result in lack of fusion or penetration and, in extreme cases, freezing of the filler into the weld pool. The addition of the filler is a result of dipping the rod into the weld pool and **not** melting off the end of the rod with the flame and allowing it to drop into the weld pool.

Ensure that when the back and forth dipping action of the filler rod takes place it is not removed from the protective outer envelope of the flame, as this may introduce impurities into the weld.

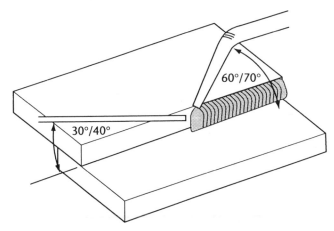

Figure 5.17 Torch and rod angles for fillet weld lap joint in flat position.

Fillet weld lap joint

The details for a fillet weld lap joint on a 3-mm thick steel plate welded in the flat position using a 2.4 mm filler rod and the leftward technique are given in Table 5.2.

Table 5.2

Plate thickness	Nozzle size	Oxygen and acetylene
3 mm	5–7	0.21 bar 3 lb/in^2
	Slope angle	Tilt angle
Torch	60–70°	50–60°
Filler	30–40°	45–50°

Ensuring close fit up between plate surfaces, tack along the joint line; use tacks 3–4 mm long at each end of the joint and at 50–60 mm spacing. Start welding at the right-hand end of the joint by forming the weld pool. The free-standing vertical edge will tend to melt slightly quicker than the bottom (flat) plate surface. Melting of both surfaces into the corner of the joint is important if full penetration into the root is required. The filler wire should be fed into the leading edge of the weld pool toward the top edge. Progress the weld along the joint at the rate of formation of the weld pool and filling out of the weld profile (see Fig. 5.17).

Fillet welded closed corner joint

The details of a fillet welded closed corner joint on a 3-mm thick steel plate, welded in the flat position using a 2.4 mm filler rod and the leftward technique are given in Table 5.3.

Table 5.3

Plate thickness	Nozzle size	Oxygen and acetylene
3 mm	5–7	0.21 bar 3 lb/in^2
	Slope angle	Tilt angle
Torch	60–70°	Equal to either side
Filler	30–40	Equal to either side

Ensuring close fit up of the joint (corner to corner), tack at each end and along the joint line at 50–60 mm spacing. Start welding at the right-hand end of the joint by forming the weld pool. The weld pool should be formed equally, to the top edge of both plates and down into the bottom of the joint. At this stage the filler rod can be added to the leading edge of the weld pool and the weld progressed along the joint at the rate of formation of the weld pool and achievement of the weld profile (see Fig. 5.18).

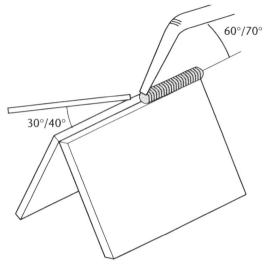

Figure 5.18 Torch and rod angles for outside corner joint in flat position.

Square edge butt joint

The details of a square edge butt joint on a 3-mm thick steel plate welded in the flat position using a 2.4 mm filler rod and the leftward technique are given in Table 5.4.

Table 5.4

Plate thickness	Nozzle size	Oxygen and acetylene
3 mm	5–7	0.21 bar
		3 lb/in^2
	Slope angle	Tilt angle
Torch	60–70°	50–60°
Filler	30–40°	45–50°

Set up the plates with a root gap approximately 1.6 mm at the right-hand end of the joint, tapering to approximately 3 mm over a 200 mm length. Tack weld at each end and at 60–70 mm spacing along the joint. Start welding at the right-hand end of the joint by forming the weld pool, ensuring melting of both edges. A slight sinking of the weld pool and a small hole forming at the leading edge will give an indication of penetration; at this stage filler can be added. If the plate edges and weld pool become too hot, remove the heat of the flame (momentarily) away from the weld pool

allowing it to cool slightly, thus maintaining control. Progress the weld at the rate of formation of the weld pool and build up of the desired weld face profile (see Fig. 5.19).

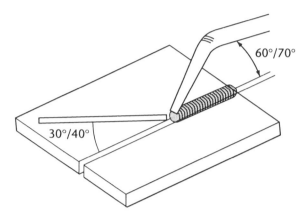

Figure 5.19 Torch and rod angles for square edge butt joint.

Horizontal–vertical 'T' fillet joint

The details of a horizontal–vertical 'T' fillet joint on a 4-mm thick plate welding in the flat position using a 2.4–3.2 mm filler and the leftward technique are shown in Table 5.5.

Table 5.5

Plate thickness	Nozzle size	Oxygen and acetylene
5 mm	10–13	0.21 bar
		3 lb/in^2
	Slope angle	Tilt angle
Torch	60–70°	45 ± 10°
Filler	20–30°	Equal to either side

Tack weld at each end and along the joint line at 80–100 mm spacing. Start the weld at the right-hand end of the joint by forming the weld pool equally on both plates and into the root of the joint. At this point the filler rod can be added to the top, leading edge of the weld pool, ensuring good fusion with the horizontal face of the joint. Pay particular attention to the deposited weld profile to avoid overlap and adhesion (rather than fusion) to the horizontal face of the joint. Progress the weld along the joint at the rate of formation of the weld pool and achievement of desired weld profile. A slight circular movement of the torch may be required but care must be taken to ensure fusion at the root of the joint (see Fig. 5.20).

pool is formed, apply the filler rod to the leading edge of the weld pool on the top edge in small amounts, allowing the molten metal to flow over the weld pool by torch manipulation. Progress along the joint at a speed consistent with weld pool development and formation of the desired weld shape (see Fig. 5.21).

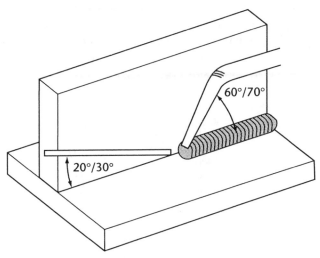

Figure 5.20 Torch and rod angles for 'T' fillet joint in the flat position.

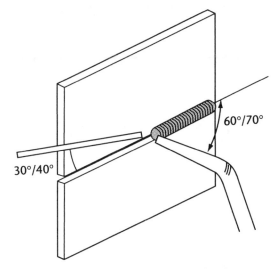

Figure 5.21 Torch and rod angles for horizontal–vertical square edge butt joint.

Square edge butt joint

Details of a square edge butt joint in the horizontal vertical position on a 3-mm thick steel plate using a 2.4 mm filler rod, a root gap of 2–3 mm and the leftward technique are given in Table 5.6.

Table 5.6

Plate thickness	Nozzle size	Oxygen and acetylene
3 mm	5–7	0.14 bar 2 lb/in^2
	Slope angle	Tilt angle
Torch	60–70°	80–90° to bottom plate
Filler	30–40°	90° ± 10°

Tack weld the plates at each end and at 70–80 mm spacing along the joint line. Start welding at the right-hand end of the joint establishing the weld pool. When the weld

Square edge butt joint

Details of a square edge butt joint on a 4-mm thick steel plate in the vertical position, welded vertical up using a 2.4–3.2 mm filler rod, a root gap of 1.6–3 mm and the leftward technique are given in Table 5.7 (also see Fig. 5.22).

After tacking at bottom and top of joint and at 75 mm spacing, start welding at the lower end of the joint by establishing the weld pool,

Table 5.7

Plate thickness	Nozzle size	Oxygen and acetylene
4 mm	7–10	0.21 bar 3 lb/in^2
	Slope angle	Tilt angle
Torch	70–80°	Equal to either side
Filler	30–40	Equal to either side

ensuring melting of both plate edges. This is indicated by the square edges taking a circular form at the leading edge of the weld pool. At this stage the filler rod can be added from the top into the leading edge of the weld pool. Manipulation of the torch across the width of the joint area will allow even deposits of the molten weld metal. Progress the weld up the joint consistent with the melting of the plate edges and achievement of the desired weld profile.

pool. Form the weld pool by melting the two edges and into the root of the joint. At this stage the filler rod can be added from the top into the leading edge of the weld pool, allowing the weld metal to fill up the joint area. Progress slowly up the joint by adding small amounts of filler wire consistent with the formation of the weld pool and melting of the plate edges. Excessive heat build-up will cause sagging of the weld metal and incorrect weld profile. Heat build-up can be controlled by momentarily removing the flame away from the weld pool (see Fig. 5.23).

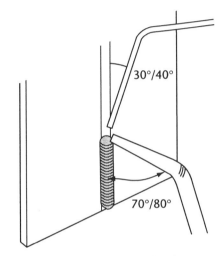

Figure 5.22 Torch and rod angles for square edge butt joint in the vertical position.

Closed corner joint

Details of a closed corner joint on a 5-mm thick steel plate in the vertical position, welded vertical up using 2.4–3.2 mm filler rod and the leftward technique are given in Table 5.8.

Tack weld the joint and start welding at the lower end of the joint by establishing the weld

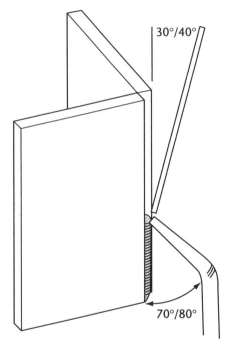

Figure 5.23 Torch and rod angles for an outside corner joint in the vertical position.

Square edge butt joint

Details of a square edge butt joint on a steel plate 5–8-mm thick welded in the flat position using 2.4–3.2 mm filler rod and the rightward technique are given in Table 5.9.

After tacking the plates with a 3–4 mm root gap, start welding at the left-hand end of the joint, the torch preceding the filler rod. Start welding by establishing the pool equally on both sides of the joint. When the weld pool is formed the filler rod can be added into the weld pool with a circular motion and the torch moving steadily forward with the formation of

Table 5.8

Plate thickness	Nozzle size	Oxygen and acetylene
5 mm	10–13	0.28 bar 4 lb/in^2
	Slope angle	Tilt angle
Torch	70–80°	Equal to either side
Filler	30–40°	Equal to either side

Table 5.9

Plate thickness	Nozzle size	Oxygen and acetylene
6–8 mm	10–25	0.28–0.42 bar 4–6 lb/in^2
	Slope angle	Tilt angle
Torch	40–50°	Equal to either side
Filler	30–40°	Equal to either side

the weld pool and weld metal build-up. It may be possible to fill the whole of the preparation in one operation; the control of the molten metal and penetration will dictate this. A second pass may be required to get the required weld profile.

The advantages of the rightward technique are less distortion, faster welding speeds, no joint preparation required up to thicknesses of 8 mm saving on filler wire. Where 'V' preparation is required, including angles are smaller than for the leftward techniques.

Even with operators skilled in the process of oxyacetylene welding, extensive practice is required to master this technique, especially when applied to positional welding.

Although the technique is still around, the development of other processes has limited its use. Where thicker sections are to be welded, consideration should be given to other processes for reasons of economy, productivity, operator fatigue, skill availability and quality of the finished weld.

Bronze welding

The oxyacetylene process can also be used for bronze welding of steel, copper, wrought and cast iron. The technique used is similar to that used for fusion, although it is **not** a fusion process. Bronze welding uses a filler which melts at a lower temperature than the base metal. Filler is deposited in the weld area and forms a bond between the fusion faces. The joint faces must be free from oil, grease, scale and rust. Suitable preparation can be achieved by wire brushing or filing; on some occasions light grinding can be used. Bronze welding produces a strong bond between the surfaces

to be joined, and is used where melting of the base metal is undesirable. This method of joining is relatively cheap and, because of the lower heat input, distortion can be reduced. An added requirement of this method of joining is the requirement of a flux which further assists in the cleaning process of the weld area.

Fluxes

Fluxes are also required when welding metals where the oxide film which forms on the surface melts at a higher temperature than the base metal. The purpose of the flux is to remove the high melting point surface oxides. Aluminium fluxes, for example, are very corrosive and must be cleaned off immediately after welding. The importance of flux removal may result in the choice of an alternative method of welding, e.g. TAGS or MAGS.

Most fluxes contain compounds which will cause irritation on skin contact, and when heated give off irritating fumes. Observe basic personal hygiene rules and ensure good ventilation when using fluxes.

The oxyacetylene process has a wide variety of applications and is used in many industrial sectors. Where problems arise, such as flux removal in the case of aluminium, or the pressure of production rates, alternative welding processes may prove more suitable.

Oxyfuel gas cutting

The oxyfuel gas cutting process provides a means of cutting steel by the chemical action of oxygen on steel. A controlled stream of oxygen is directed onto the surface of the steel which has been pre-heated to 900–950°C (ignition temperature). The result of this action is that the pre-heated area rapidly becomes oxidised and the oxide is blown away by the oxygen stream severing the plates. The width of the cut being referred to as the **kerf** width.

The equipment used for oxyfuel gas cutting will concentrate on the supply of gases in high-pressure cylinders as discussed earlier. For this reason the safety precautions relating to the use of gas cylinders, components in the system and their function, assembly of equipment and testing, will equally apply.

The main difference is in the torch which is used for cutting – see Fig. 5.24.

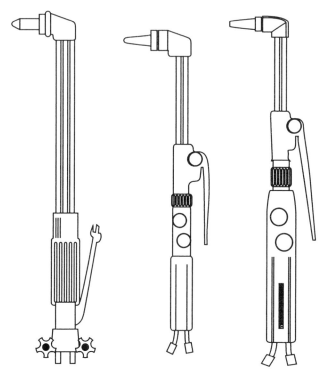

Figure 5.24 Oxyfuel cutting torches.

Torches

The torch provides a means of mixing the gases either in the body of the torch or in the nozzle, to provide a pre-heating flame and a separate supply of oxygen for cutting. In the case of the nozzle mix types the pre-heating oxygen and acetylene mix takes place just prior to the gases being burned at the nozzle end (see Fig. 5.25).

The two main functions of the pre-heat flame are:

1 It raises the starting point of the cut to the ignition temperature prior to cutting, and provides a continuous heat source to enable the progression of the cut and the continuation of the chemical action of the oxygen at the plate surface, and

2 The action of the preheat flame on the surface of the plate will remove any contami-

nation or applied surface preparations such as paint. Where plate surfaces are heavily contaminated, surface cleaning is recommended (scraping or light grinding) prior to cutting.

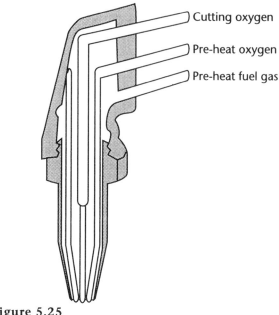

Figure 5.25

Fuel gases

Possibly the two most common forms of fuel gas used for cutting are acetylene and propane. Both gases produce good-quality cuts, but for general-purpose use acetylene is the most versatile and economical of the two. Table 5.10 shows a comparison.

Table 5.10

Operation	Acetylene	Propane	Apachi
Starting cut	✔✔✔	✔	✔✔
Cutting speed	✔✔✔	✔✔	✔✔✔
Cost	✔	✔✔✔	✔✔
Bevelling	✔✔✔	✔	✔✔
Hole piercing	✔✔✔	✔✔✔	✔✔
Thickness	✔✔✔	✔	✔✔

✔✔✔, Best choice.

Nozzles

The nozzle used with an oxygen and acetylene mix has an outer ring of holes which provides the pre-heat flame with a single central hole for the cutting oxygen supply. This nozzle is made of copper and comes in one piece (see Fig. 5.26).

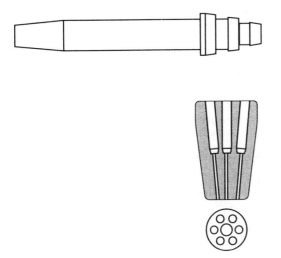

Figure 5.26 Oxygen/acetylene nozzle.

The propane nozzle is usually made up of two parts, a brass inner part – fluted (grooved) around the outside providing the pre-heat flame with a central hole for the cutting oxygen, and an outer copper surround (see Fig. 5.27).

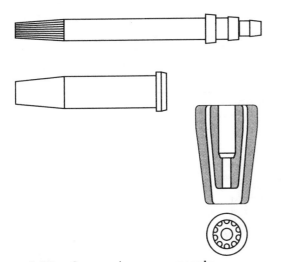

Figure 5.27 Oxygen/propane nozzle.

The nozzle for Apachi gas is made up of two parts; an 18-fluted (grooved) brass inner and a copper outer, designed to fit a variety of cutting torches.

Maintaining the nozzle in good condition is the answer to achieving a good cut. If the nozzle is damaged on the end, or if any of the holes become blocked, the quality of the cut will suffer. The one-piece acetylene nozzle can be cleaned by light abrasion with emery cloth on a flat surface, ensuring that the nozzle is kept square to the abrasive surface. Special reamers (nozzle cleaners) are available which match the hole sizes in the nozzle. By inserting the correct size cleaner into the hole using a twisting motion, the hole can be effectively freed of obstruction.

The condition of the nozzle will also affect the kerf width (width of cut). Contamination or damage to the pre-heating holes will give uneven or insufficient pre-heating of the plate surface, resulting in poor-quality cuts. Similarly when the central hole is damaged, misshapen or contaminated, the cutting oxygen stream will be distorted – again resulting in poor quality cuts.

If the plate thickness increases, the size of the cutting nozzle will increase as indicated in Table 5.11. The measurement given in inches corresponds to the size of the cutting oxygen hole.

Cutting data for acetylene and propane are give in Tables 5.12 and 5.13, respectively.

Table 5.11

Size (mm)	(in)	Plate thickness (mm)	(in)
3–6	1/32	6	1/4
5–12	3/64	12	1/2
10–75	1/16	75	1–3
70–100	5/64	100	4
90–150	3/32	150	6
190–300	1/8	300	12

Hand cutting

So far the emphasis on achieving quality cut edges has been dependent on the condition of the equipment and the correct gas pressures. This must also be supported by an even speed of travel and keeping a constant distance between the end of the nozzle and the plate surface. When hand cutting, considerable practice is required to master this skill; for this reason the use of a cutting guide is recommended. This may consist of a roller-type guide attached to the nozzle which rests on the plate surface or straight edge, helping in keeping the nozzle-to-work distance constant (see Fig. 5.28)

Table 5.12

Plate thickness		Nozzle	Operating pressure				Gas consumption					
			Oxygen		Fuel		Cutting Oxygen		Heating Oxygen		Fuel	
mm	in	size	bar	lbf/in²	bar	lbf/in²	l/h	ft³/h	l/h	ft³/h	l/h	ft³/h
6	1/4	1/32	1.8	25	0.14	2	1000	28	480	15	400	14
13	1/2	3/64	2.1	30	0.21	3	1900	67	57	20	510	18
25	1	1/16	2.8	40	0.14	2	4000	140	640	19	470	17
50	2	1/16	3.2/3.5	45/50	0.14	2	4500	160	820	22	560	19
75	3	1/16	3.5/4.8	50/60	0.14	2	4800	170	680	24	620	22
100	4	5/64	3.5/5.5	45/70	0.14	2	6800	240	850	30	790	27
150	6	3/32	4.2	45/80	0.21	3	9400	330	960	34	850	30
200	8	1/8	5.3	60	0.28	4	14800	510	1380	48	1250	44
250	10	1/8	6.3	75	0.28	4	21500	760	1580	55	1420	50
300	12	1/86.7	6.7	90	0.28	4	25000	680	1580	55	1420	50
Sheet		Asnm	1.5	20	0.14	2	800	206	85	3	85	3

Table 5.13

Plate thickness		Nozzle	Operating pressure				Gas consumption						App Cutting	
			Oxygen		Fuel		Cutting Oxygen		Heating Oxygen			Fuel	Speeds	
mm	in	size	bar	lbf/in²	bar	lbf/in²	l/h	ft³/h	l/h	ft³/h	l/h	ft³/h	mm.m	in.m
6	1/4	1/32	2.1	30	0.2	3	1000	36	1300	48	300	12	430	17
13	1/2	3/64	2.12	30	0.2	3	1800	65	1600	57	300	14	360	14
25	1	1/16	2.8	40	0.2	3	3000	140	1700	62	400	15	280	11
50	2	1/16	3.2	45	0.3	4	4500	160	1800	66	400	16	205	8
75	3	1/16	3.5	50	0.3	4	4800	170	2000	73	500	18	205	8
100	4	5/64	3.5	50	0.3	4	7300	260	2600	93	600	23	152	6
150	6	3/32	4.2	60	0.4	6	12300	435	3300	120	800	30	125	5
250	10	1/8	5.6	80	0.6	8	22300	790	4600	165	1100	42	60	2
300	12	1/86.7	6.7	95	0.6	8	26300	930	5900	210	1400	50	50	2

Data are for guidance only and may vary with operating conditions, materials, etc.
Gas pressures are shown in bar – 1 bar = 1 kg cm²; lbfin² = 0069 bar.
Gas consumption in litres per hour (l/h).

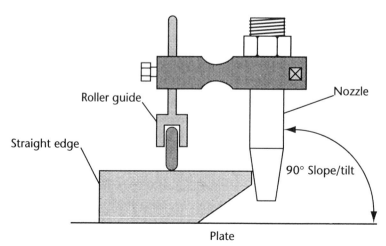

Figure 5.28   Cutting guide.

The nozzle must be maintained at 90° to the plate surface (slope and tilt). Any variation either side will reflect on the cut resulting in the plate edge not being square to the surface. Other attachments are available for circle cutting; two examples are shown in Fig. 5.29.

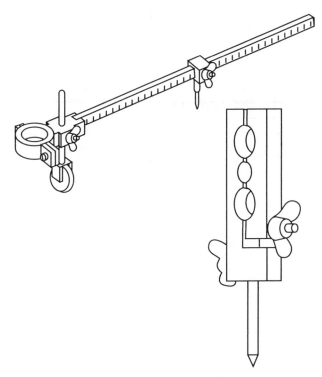

Figure 5.29   Attachments for circle cutting.

All of the above will influence the achievement of accurate quality cuts. Better control of maintaining consistent cutting speeds, nozzle-to-work distance and on straight line is by the use of cutting machines (straight line cutter). With this type of machine the cutting head is mounted on a motorised unit with a cutting speed control and adjustment of the nozzle-to-work distances. The whole unit runs on a track to maintain a straight line. Other types of cutting machines are available which can support a number of cutting heads and for accurate cutting of complicated and intricate shapes, e.g. profile cutters.

Quality of cut

Examination of an edge will show a sequence of vertical lines (drag lines). The deflection of these lines from the vertical plane will give an indication of the speed of travel, nozzle-to-work distance, and the condition of the pre-heat flame (see Fig. 5.30).

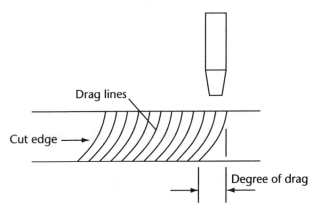

Figure 5.30   Degree of drag.

The edge of a good cut will display the following features:

a) vertical drag lines
b) bottom edge free from attached slag
c) square face
d) easily removable scale (with file)
e) square corners at top and bottom of cut
f) drag lines just visible.

If the pre-heat flame temperature is too low:

a) delay in achieving cut temperature
b) reduced cutting speeds
c) distorted drag lines
d) gouging effect on bottom edge of cut face
e) unacceptable cuts
f) continuous loss of cutting action.

If the pre-heat flame temperature is too high:

a) excessive melting of top edge
b) adhesion of cut faces by melting of top edge
c) excessive slag adhesion to bottom edge
d) distorted drag lines
e) tapering of cut face from top to bottom.

Speed of travel too fast:

a) loss of cutting action
b) excessive drag, opposite direction of travel
c) misshapen top edge of cut
d) irregular cut face
e) rounded bottom edge of cut preventing drop cuts.

If the nozzle-to-work distance is too high

a) excessive rounding of top edge
b) excessive melting of top edge
c) square bottom edge
d) lower part of cut may be square
e) gouging at top edge.

If the nozzle-to-work distance is too low

a) square cut face
b) square bottom edge
c) slightly rounded top edge
d) possible heavy beading on top edge.

If the nozzle is contaminated

a) excessively wide kerf width
b) pleated appearance of top edge
c) rounding of top edge
d) irregular bottom edge
e) slag adhesion on bottom edge
f) rough, irregular cut face.

If the speed of travel is too slow

a) excessive rounding on top edge of cut
b) heavy scale on cut face
c) distorted drag lines
d) irregular bottom edge of cut
e) excessive slag adhesion on cut face.

Observing a few basic rules when oxyfuel gas cutting will help in making good-quality cuts and remove the need for further or excessive welding preparation:

1 Ensure plate surfaces are clean and free from scale.
2 Select the correct nozzle for the job in hand.
3 Ensure the nozzle is free of obstruction, pre-heat and oxygen.
4 Select the correct operating pressures.
5 Use cutting attachments to maintain i) the nozzle-to-work distance and ii) the nozzle 90° to the plate surface in both planes.
6 Correct and consistent speed of travel.

Supporting materials

Where it is practical to do so, the metal to be cut should be supported to make sure that cutting can be done safely, without damage to underlying components. There are many different types of support and vary from open lattice top bunches to individual adjustable

supports. Upturned channel sections or angle iron provides a suitable support and allows cuts to be made without hampering the cutting process (see Fig. 5.31).

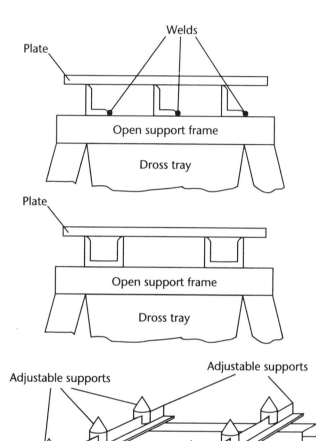

Figure 5.31   Work support.

Cutting techniques

Having followed the 'equipment assembly' procedure covered earlier, select the correct operating pressure and nozzle size for the job in hand. A neutral flame is required for oxyacetylene cutting, this can be achieved by opening the acetylene control valve on the torch and lighting with a spark lighter. The acetylene flame should burn without excessive smoke or sooty depsosits. Slowly open the oxygen valve until the blue cones of the preheating flame are clearly defined.

Starting from a plate edge: position the nozzle over the point where the cut is to start so that the central hole of the nozzle is on line with the plate edge and the ends of the preheat cones just clear of the plate surface (see Fig. 5.32).

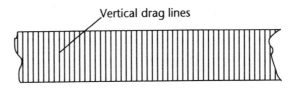

Figure 5.33   Drag lines on good-quality cut.

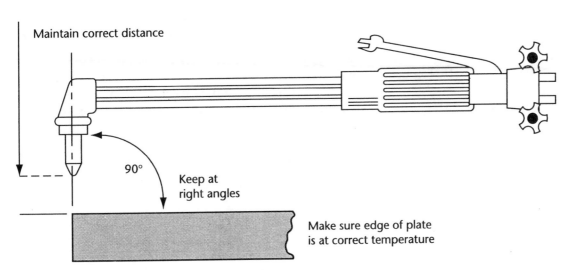

Figure 5.32   Conditions for good-quality cut.

Hold the cutting torch at this point until the plate edge is bright red in colour. At this point fully depress the cutting oxygen lever allowing oxygen stream to flow from the central hole, at the same time pulling (or pushing) the cutting torch along the cutting line. The speed of travel will be consistent with the continuous heating of the leading point of the cut. Continue the process maintaining the nozzle-to-work distances and a consistent speed of travel over the required length of cut. If the correct conditions have been achieved the cut portion should drop away without adhesion, leaving a light oxide on the cut face. The top and bottom edge of the cut should be square with slightly noticeable vertical drag lines (see Fig. 5.33).

Starting inside the plate (hole piercing): start by holding the torch stationary over the point at which the cut is to start. When the area has reached the bright red colour, slightly raise the torch before fully depressing the cutting oxygen level. On depressing the cutting oxygen level, a small amount of metal may bubble up onto the plate surface while the stream of oxygen pierces the plate. When the hole is established, lower the cutting torch to the original position and start the cut as before. Figure 5.34 shows the correct sequence: (1) preheat the plate surface, (2) raise the nozzle slightly and depress cutting oxygen lever piercing the hole, and (3) lower the nozzle to the original position and start the cut.

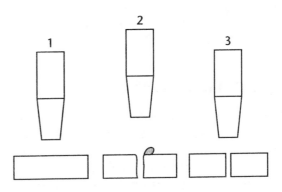

Figure 5.34

Starting the cut on a round face (e.g. round bar)

Starting cuts on the circumference of a round bar, although possible, can be difficult. To overcome this use a diamond point or round nose chisel and raise a burr at the point at which the cut is to start. Concentrating the pre-heat flame and the cutting oxygen stream at the point will enable an easier, quicker start to the cutting operation (see Fig. 5.35).

Applying the above will provide a basis for successful cutting of steels. Experience will be gained with practice.

In the interests of safety it is strongly advised not to tamper with the equipment used for oxyacetylene cutting or welding. Never attempt to repair the equipment yourself. Manufacturers and suppliers provide a

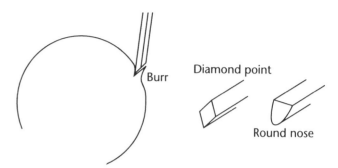

Figure 5.35   Starting the cut on a round surface.

service to part-exchange most of the equipment used, and supply exchange components which have been serviced to rigid codes and standards. If in doubt about any aspect of the process or equipment, it is advisable to seek the manufacturer's or supplier's advice.

6

Questions

Health and safety

The Health and Safety at Work Act 1974 states that an employee (welder) is responsible for

A) safety of himself only
B) safety of himself and others
C) wearing of ear plugs at all times
D) providing his own protective equipment.

MMA welding

A1 The purpose of a cover glass in a head-shield is to protect the filter lens from
A) radiation
B) infra-red rays
C) light
D) spatter.

A2 A welding earth is connected between ground and
A) work
B) welding power unit
C) mains
D) return lead.

A3 An electrode size is identified by the
A) type of flux coating
B) diameter of the core wire
C) length of the core wire
D) diameter of the flux coating.

A4 During manual metal-arc welding, the electrode will be consumed faster if
A) the welding current is decreased
B) the speed of travel is increased
C) the welding current is increased
D) a weaving technique is used.

A5 The maximum open circuit voltage allowed for manual metal-arc welding is
A) 100 V
B) 220 V
C) 330 V
D) 440 V.

A6 Two pieces of low carbon steel of 8 mm thickness are to be butt welded together in the flat position using conventional electrodes. The correct joint preparation would be
A) closed square butt
B) single vee 30° included angle
C) open square butt
D) single vee 60° included angle.

A7 A root bend test is applied to a butt welded joint in low carbon steel. If the specimen did not fracture, this would indicate
A) correct weld face contour
B) absence of undercut
C) good root fusion
D) lack of fusion.

A8 A respirator is used to protect the welder from
A) electric shock
B) harmful fumes
C) infra-red rays
D) hot metal particles.

A9 A satisfactory permanent method of connecting return lead clamps to workbenches is by
A) soldering
B) brazing
C) bolting
D) clamps.

A10 It is an advantage to carry out manual metal-arc welding in the flat position rather than the horizontal–vertical position because the electrodes used may be
A) bare wire
B) shorter
C) larger diameter
D) smaller diameter.

B11 State the purpose of a fuse in an electric circuit.

B12 Name TWO welding power units which supply direct current to a welding arc.

B13 State TWO reasons why electrodes for manual metal-arc welding need careful storage.

B14 State TWO reasons why the correct arc length should be maintained during manual metal-arc welding.

B15 Give ONE reason why it is essential that correct joint preparations are used when manual metal-arc welding.

B16 Give TWO reasons for using the multi-run technique when producing a horizontal-vertical tee-fillet weld.

B17 Give TWO reasons why slag should be removed before depositing the next run of weld metal.

B18 Name TWO items of protection to be worn by the welder during manual metal-arc welding operations.

B19 State ONE cause of a bolted electrical cable connection overheating during use.

B20 State TWO items of information which may be obtained from a macro-examination of a welded joint.

MAGS welding

A1 The cable connecting the work to the power source unit is called the welding
A) earth
B) mains
C) return
D) output.

A2 The glass tube situated adjacent to the regulator of a MAGS unit is a
A) contents gauge
B) heater unit
C) filter unit
D) flowmeter.

A3 When using liquid CO_2 for the MAGS welding process, it is necessary to incorporate into the system a
A) filter
B) heater
C) cooler
D) by-pass valve.

A4 A cylinder painted black with a vertical white line contains
A) oxygen gas
B) CO_2 gas
C) CO_2 liquid
D) argon gas.

A5 The gas supplied in blue cylinders with a green band contains a mixture of argon and
A) oxygen
B) helium
C) hydrogen
D) CO_2.

A6 The filler wire used in MAGS welding is fed to the weld pool from a
A) spool
B) heater
C) pool
D) solenoid.

A7 The slope angle of a MAGS gun when making a bead in the flat position should be
A) 30°
B) 40°
C) 50°
D) 70°

A8 A useful ancillary item of equipment necessary when MAGS welding is
A) a spark lighter
B) a powdered flux
C) wire cutters
D) nozzle cleaners.

A9 The abbreviation o.c.v. stands for
A) outlet current voltage
B) open current voltage
C) outlet circuit voltage
D) open circuit voltage.

A10 The recommended filter shade number for MAGS welding up to 200 A is
A) 6 E/W
B) 8 E/W
C) 10 E/W
D) 12 E/W.

B11 Name the angle indicated at 'A' on the figure below.

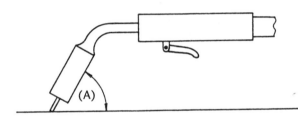

B12 State two reasons why the steel filler wire used in the MAGS welding process is copper coated.

B13 Sketch a contact tip as used in a MAGS welding gun.

B14 The following figure shows a simple line diagram of a wire feed mechanism. Name the parts numbered 1, 2, 3 and 4.

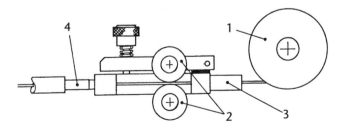

B15 Name four hazards associated with arc welding processes.

B16 Name one destructive test used in welding practice.

B17 Name the two defects shown in the figure.

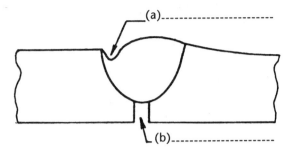

B18 Name the mode of metal transfer used in the MAGS welding process for welding thinner sections and positional work.

B19 Name and sketch the recommended joint preparation for butt welding 2 mm thick low carbon steel in the flat position by the MAGS process.

B20 State one reason why arc welding shields are designed for full face protection.

TAGS welding

A1 Argon gas is dangerous when used in confined spaces because it may cause
A) defects
B) enrichment
C) poisoning
D) fumes.

A2 The two weld defects shown in the figure below are
A) undercut and porosity
B) porosity and cracks
C) undercut and cracks
D) porosity and slag.

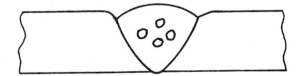

A3 The nozzle used to direct the shielding gas onto the weld area in TAGS welding is made from
A) aluminium
B) plastic
C) steel
D) ceramic.

A4 The TAGS weld shown in the figure is an outside
A) angle
B) triangle
C) corner
D) filler.

A5 The pressure contained in a full cylinder of argon gas is
A) 2 bars
B) 20 bars
C) 200 bars
D) 2000 bars.

A6 A typical gas flow rate through a TAGS torch, in litres per minute, may be
A) 8
B) 80
C) 800
D) 8000

A7 In TAGS welding, an accurate flow of shielding gas is ensured by combining a flowmeter with a
A) reformer
B) reducer
C) receiver
D) regulator.

A8 For safety reasons, every TAGS welding power source must be fitted with an electrical
A) isolator
B) activator
C) regulator
D) inductor.

A9 A TAGS welding area should be adequately protected from
A) dilution
B) draughts
C) dispersion
D) diffusion.

A10 Macro-etch testing uses an etchant which may be a mixture of acid and
A) wax
B) acetone
C) alkali
D) water.

B11 Sketch in good proportion the shape of a tungsten electrode ground for use in a d.c. circuit.

B12 Produce two sketches to show the following joints:
A) tee fillet, flat position
B) tee fillet, horizontal vertical position.

B13 State one condition under which TAGS filler wire should be stored.

B14 Show on the figure the approximate angle of slope for a TAGS torch.

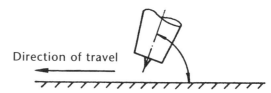

Direction of travel

B15 Name two items of safety equipment or clothing suitable for TAGS welding.

B16 Sketch a butt weld indicating excess penetration.

B17 State the purpose for which a rectifier is used.

B18 Show, by means of a sketch, the form of an open square butt joint.

B19 A tungsten electrode, connected to the negative pole of a d.c. power source, melts during use. State a probable cause.

B20 State the diameter of any two standard TAGS filler wires.

Gas welding and cutting

A1 The one-way component fitted between the hoses and the blowpipe in an oxyacetylene welding system is called a hose
A) connector
B) reducer
C) regulator
D) protector.

A2 Oxygen is supplied in cylinders which are painted
A) blue
B) red
C) maroon
D) black.

A3 The flame setting used for fusion welding low carbon steel by the oxy-acetylene process is
A) oxidizing
B) carburizing
C) neutral
D) normalizing.

A4 Acetylene fittings have threads which are
A) left hand
B) right hand
C) square
D) round.

A5 The oxygen hose on a blowpipe for oxy-acetylene welding is coloured
A) red
B) blue
C) green
D) grey.

A6 Gas pressures used in oxy-acetylene welding are measured in
A) newtons
B) bars
C) grams
D) tonnes.

A7 The angle of filler wire to the work when oxy-acetylene welding by the leftward technique in the flat position is in the range of
A) 10/20°
B) 20/30°
C) 30/40°
D) 60/70°

A8 The angle of the nozzle to the work when oxy-acetylene welding by the leftward technique in the flat position is in the range of
A) 10/20°
B) 20/30°
C) 30/40°
D) 60/70°

A9 The size of filler wire diameter required for welding 1.6 mm low carbon steel is
A) 1.6 mm
B) 2.4 mm
C) 3.2 mm
D) 4.0 mm

A10 A macro-examination of a butt weld involves polishing and
A) etching
B) bending
C) heating
D) fracturing.

B11 Sketch the end face of an oxy-acetylene cutting nozzle and indicate the orifice through which the cutting stream of oxygen is emitted.

B12 State two functions of a flashback arrester.

B13 State why copper must never be used on an acetylene supply.

B14 State a safe method generally used for testing for leaks on oxy-fuel gas welding or cutting systems.

B15 State why acetylene cylinders must always be stored and used standing in the vertical position.

B16 Sketch a neutral oxy-acetylene flame.

B17 On the sketch of the acetylene cylinder shown in Fig. 6.7, indicate the position of
A) the safety valve
B) the fusible plug.

B18 State the main advantage of a two-stage regulator compared with a single-stage regulator.

B19 Name two weld defects.

B20 State two hazards associated with oxy-acetylene welding.

Answers to questions

Health and safety

B).

MMA welding

A1 D   A6   D
A2 A   A7   C
A3 B   A8   B
A4 C   A9   C
A5 A   A10 C   *[1 mark each]*

B11 To protect the circuit from an overload of current.
                                    [2 marks]

B12 DC generator; rectifier.   *[1 mark each]*

B13 To keep them dry; To prevent damage; To avoid mixing types and sizes.
                          [Any two; 1 mark each]

B14 To prevent atmospheric contamination; To reduce spatter losses; To maintain correct power level.
                          [Any two; 1 mark each]

B15 To ensure correct penetration; satisfactory fusion; economy of materials
                          [Any one; 2 marks]

B16 To ensure the correct weld form; To allow the use of smaller electrodes for better control of weld pool.*[1 mark each]*

B17 To ensure correct fusion; To avoid slag inclusions; To permit ease of control of weld pool.   *[Any two; 1 mark each]*

B18 Apron, cape, leggings, spats, gloves, goggles, face shields. *[Any two; 1 mark each]*

B19 The connection is loose; Frayed cable end.   *[Any one; 2 marks]*

B20 Weld defects; Heat affected zone; Disposition of runs.
                          [Any two; 1 mark each]

MAGS welding

A1 C   A6   A
A2 D   A7   D

A3 B   A8   C
A4 C   A9   D
A5 D   A10 C   *[1 mark each]*

B11 Slope angle – (angle given in degrees not acceptable).   *[2 marks]*

B12 Improve electrical conductivity; Reduce friction; Minimise corrosion. *[Any two; 2 marks]*

B13   *[2 marks]*

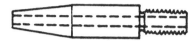

B14 1 Spool of wire; 2 wire feed rolls; 3 inlet guide; 4 outlet guide.
                          [$^1/_2$ mark each; 2 marks]

B15 Fumes, burns, sparks, radiation, hot metal, electric shock, enrichment of atmosphere, etc., and any other relevant
                  [$^1/_2$ mark each × 4 = 2 marks]

B16 (a) Macro-etching; (b) Bend test.
                  [2 marks for either answer]

B17 (a) Undercut; (b) Lack of penetration.
                          [1 mark each]

B18 Dip
                                    [2 marks]

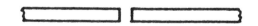

B19
        Drawing *[1 mark]*; Square butt *[1 mark]*.
                                    [2 marks]

B20 To protect from harmful arc emissions, e.g. ultra violet, infra red or heat.
                                    [2 marks]

TAGS welding

A1 B   A6   A
A2 A   A7   D
A3 D   A8   A
A4 C   A9   B
A5 D   A10 D   *[1 mark each]*

B11 [*2 marks*]

B12 [*1 mark each*]

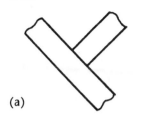 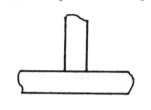

(a)   (b)

B13 In dry, warm conditions; On correctly labelled racks; Away from dampness or chemicals; Away from oil, grease, dirt, etc. [*Any one; 2 marks*]

B14 [*2 marks*]

70–80°

Direction of travel

B15 Gloves, screen, overalls, apron, safety boots, cap, etc.

[*Any two; 1 mark each*]

B16

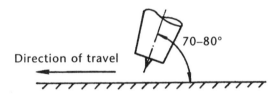

B17 To change a.c. to d.c. [*2 marks*]

B18 Gap to be at least between *T* to 1/2*T*.

[*Shape 1 mark; Gap 1 mark*]

B19 The current is too high; Electrode is contaminated.

[*Any one; 2 marks*]

B29 1.0 mm; 1.6 mm; 2.4 mm; 3.2 mm; 4.0 mm; 5.0 mm

[*Any two; 2 marks*]

Gas welding and cutting

A1	D	**A6**	B
A2	D	**A7**	B
A3	C	**A8**	D
A4	A	**A9**	A
A5	B	**A10**	A

[*1 mark each*]

B11 One mark for nozzle end face as shown (accept other numbers of peripheral holes); 1 mark for indicating cutting stream orifice.

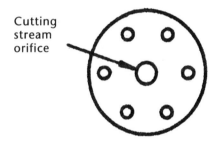

Cutting stream orifice

B12 (a) To prevent any burning gases getting back to the gas storage area, i.e. cylinders, etc. (quenching flame); (b) To prevent back pressure to cylinders or gas supply. [*1 mark each*]

B13 Forms a dangerous (or explosive) compound.

[*2 marks*]

B14 Pressurise the system; Brush each connection with a soapy water or 'teepol' solution.

[*1 mark each*]

B15 To prevent acetone being drawn out of the cylinder valve. [*2 marks*]

B16 One mark for general drawing; 1 mark for rounded inner cone.

B17 One mark for correct position of safety valve; 1 mark for correct position of fusible plug.

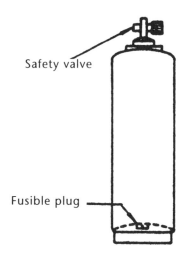

Safety valve

Fusible plug

B18 Gives a more stable outlet pressure to the blowpipe. *[2 marks]*

B19 Lack of penetration, excessive penetration, surface defects cracks, lack of fusion, inclusions, porosity or any other relevant defect (BS 499).
[Any two; 1 mark each]

B20 Fumes, burns, radiation, hot metal, welding in confined spaces, oxygen enrichment, combustible gases and gas mixtures, backfires and flashbacks.
[Any two, 1 mark each]

— Appendix 1 —
Welding processes

Welding processes and their numerical representation

1 **Arc welding**
11 Metal arc welding without gas protection
111 Metal arc welding with covered electrode
112 Gravity arc welding with covered electrode
113 Bare wire metal arc welding
114 Flux cored metal arc welding
115 Coated wire metal arc welding
118 Firecracker welding
12 Submerged arc welding
121 Submerged arc welding with wire electrode
122 Submerged metal arc welding with strip electrode
13 Gas shielded metal arc welding
131 MIG welding
135 MAG welding: metal arc welding with non-inert gas shield
14 Gas-shielded welding with non-consumable electrode
141 TIG welding
149 Atomic-hydrogen welding
15 Plasma arc welding
18 Other arc welding processes
181 Carbon arc welding
185 Rotating arc welding
915 Salt bath brazing

2 **Resistance welding**
21 Spot welding
22 Seam welding
221 Lap seam welding
225 Seam welding with strip

23 *Projection welding*
24 *Flash welding*
25 *Resistance butt welding*
29 Other resistance welding processes
29 HF resistance welding
943 Furnace soldering

3 **Gas welding**
31 Oxyfuel gas welding
311 Oxyacetylene welding
312 Oxypropane welding
313 Oxyhydrogen welding
32 Air–fuel gas welding
321 Air–acetylene welding
97 Braze welding
953 Friction soldering

4 **Solid phase welding; pressure welding**
41 Ultrasonic welding
42 Friction welding
43 Force welding
44 Welding by high mechanical energy
45 Diffusion welding
47 Gas pressure welding
48 Cold welding

7 **Other welding processes**
71 Thermit welding
72 Electroslag welding
73 Electrogas welding
74 Induction welding
75 Light radiation welding
751 Laser welding
752 Arc image welding
753 Infrared welding
76 Electron beam welding
78 Stud welding

781   Arc stud welding
 78   Resistance stud welding
  9   **Brazing, soldering and braze welding**
 91   *Brazing*
911   Infrared brazing
912   *Flame brazing*
913   Furnace brazing
914   Dip brazing
916   Induction brazing
917   Ultrasonic brazing
918   Resistance brazing
919   Diffusion brazing
923   Vacuum brazing
924   Vacuum brazing
 93   Other brazing processes
 94   Soldering

941   Infrared soldering
942   Flame soldering
944   Dip soldering
945   Salt bath soldering
946   Induction soldering
947   Ultrasonic soldering
948   Resistance soldering
949   Diffusion soldering
951   Flow soldering
952   Soldering with soldering iron
954   Vacuum soldering
 96   Other soldering processes
322   Air–propane welding
971   Gas braze welding
 72   Arc braze welding
441   Explosive welding

— Appendix 2 —

Useful information

Chemical symbols

Al	Aluminium	Nb	Niobium
C	Carbon	Ni	Nickel
Cb	Columbium	O	Oxygen
	(Niobium)	P	Phosphorus
Co	Cobalt	Pb	Lead
Cr	Chromium	S	Sulphur
Cu	Copper	Si	Silicon
H	Hydrogen	Sn	Tin
Fe	Iron	Ta	Tantalum
Mg	Magnesium	Ti	Titanium
Mn	Manganese	V	Vanadium
Mo	Molybdenum	W	Tungsten
N	Nitrogen	Zn	Zinc

Electrode diameters – metric, imperial and approximate lengths

Diameters			Lengths	
mm	SWG	in	mm	in
1.6	16	1/16	250	10
2	14	5/64	300	12
2.5	12	3/32	350	14
3.25	10	1/8	400	16
4	8	5/32	450	18
5	6	3/16	600	24
6	4	1/4	700	28
8	–	5/16		

Length

	m	cm	in	ft	yd
1 metre	1	100	39.3701	3.28084	1.0936
1 centimetre	0.01	1	0.393701	0.0328084	0.0109361
1 inch	0.0254	2.54	1	0.0833333	0.0277778
1 foot	0.3048	30.48	12	1	0.3333333
1 yard	0.9144	91.44	36	3	1

	km	mi	n.mi
1 kilometre	1	0.621371	0.539957
1 mile	1.60934	1	0.868976
1 nautical mile	1.85200	1.15078	1
Bar	Atmosphere	lb/in^2	
1	0.986923	14.6959	

Weight

	1 grain	64.8 mg
	1 dram	1.772 g
16 drams	1 ounce	28.35 g
16 oz	1 pound	0.4356 kg
14 pounds	1 stone	6.35 kg
2 stones	1 quarter	12.7 kg
4 quarters	1 hundredweight	50.8 kg
20 cwt	1 (long) ton	1.016 tonnes
	1 milligram	0.015 grain
10 mg	1 centigram	0.154 grain
10 cg	1 decimgram	1.543 grain
10 dg	1 gram	15.43 grain
0.035 oz		
1000 g	1 kilogram	2.205 lb
1000 kg	1 tonne (metric ton)	0.984 (long) ton

Area

1 are	100 m^2	119.6 yd^2
1 hectare	100 are	2.471 acres
1 km^2	100 hectares	0.387 mi^2
1 acre	0.4047 hectare	4840 yd^2
1 rood	1011.7 m^2	^{1}/4 acre
1 mi^2	2.59 km^2	640 acres

Cubic measure

	1 cubic inch	16.4 cm^3
1728 cu in	1 cubic foot	0.0283 m^3
27 cu ft	1 cubic yard	0.765 m^3
	1 cu centimetre	0.061 in^3
1000 cu cm	1 cu decimetre	0.035 ft^3
1000 cu dm	1 cu metre	1.308 yd^2

Capacity measure

	1 fluid ounce	28.4 ml
5 fl oz	1 gill	0.142 l
4 gill	1 pint	0.568 l
2 pt	1 quart	1.136 l
4 qt	1 gallon	4.546 l
	1 millilitre	0.002 pt
10 ml	1 centilitre	0.018 pt
10 cl	1 decilitre	0.176 pt
10 dl	1 litre	1.76 pt

Units and metric conversion factors

Abbreviations

m	metre
g	gramme
t	tonne (or metric ton)
N	newton (SI unit of force)
J	joule (SI unit of energy)
W	watt (SI unit of power)
s	second
M	mega (× 1 million)
k	kilo (× 1 thousand)
c	centi (1 hundredth)
m	milli (1 thousandth)
μ	micro (1 millionth)

Force

1 lbf	4.448 N
1 tonf	9.964 kN
1 kgf	9.807 N
1 N	0.2248 lbf

Conversions between lbf, kgf, etc. are as for mass units.

Pressure or stress

1 lbf/in^2	0.0703 kgf/cm^2
	6.895 kN/m^2
1 tonf/in^2	1.575 kgf/mm^2
	15.444 MN/m^2
1 kgf/cm^2	14.223 lbf/in^2
	98.067 kN/m^2
1 kN/m^2	0.145 lbf/in^2
1 MN/m^2	0.06475 ton/in^2

Energy (work, heat)

1 ft lbf	0.1383 kgf m 1.356 J
1 kgf m	7.233 ft lbf
	9.81 J
1 kW h	3412 Btu 3.6 MJ
1 Btu	1.055 kJ
1 J	0.102 kgf m/s
	1 J/sl 1 N m/s
1 kW	1.341 hp

Basic conversion factors

To convert	into	Multiply by
in	mm	25.40
mm	in	0.0393 701
ft	m	0.304 8
m	ft	3.280 839 8
lb	kg	0.453 592 370
kg	lb	2.204 62
ton (long)	tonne	1.016 05
tonne	kg	1000.0
gallon (imp.)	l (litre)	4.545.96
l	ml	1000.0
ml	cm^{-1}	1.000 028
cu. ft.	1	28.316 1

Compound conversion factors

$tonf/in^2$	N/mm^2	15.444 3
lbf/in^2	N/mm^2	006 894 777
N/mm^2	$tonf/in^2$	0.064 749
N/mm^2	lbf/in^2	145.037 76
ft.lbf.	J(joules)	1.355 82
kgf.m	J	9.806 650
kgf.m	ft.lbf.	7.233 01
ft.lbf.	kgf.m	0.138 255
J	ft.lbf.	0.737 562
in/min	m/hr	1.524 0
m/hr	in/min	0.656 168
cu.ft/hr	1/min	0.471 95
1/min	cu.ft/hr	2.118 936
lb/cu.ft.	g/cm^3	0.016 02
g/cm^3	lb/cu.ft.	62.43

Temperature

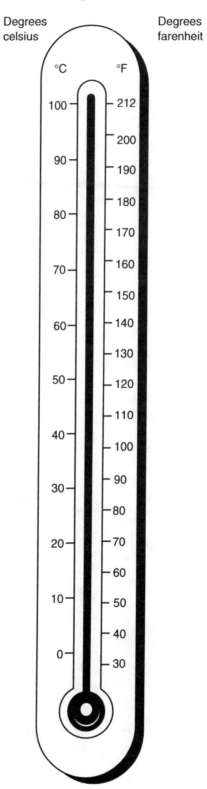

Degrees celsius

Degrees farenheit

Index